Astronomers' Universe

W0079970

For further volumes:
http://www.springer.com/series/6960

David A.J. Seargent

Weird Weather

Tales of Astronomical and Atmospheric Anomalies

 Springer

David A.J. Seargent
The Entrance
NSW, Australia

ISSN 1614-659X
ISBN 978-1-4614-3069-8   e-ISBN 978-1-4614-3070-4
DOI 10.1007/978-1-4614-3070-4
Springer New York Dordrecht Heidelberg London

Library of Congress Control Number: 2012932204

© Springer Science+Business Media New York 2012
This work is subject to copyright. All rights are reserved by the Publisher, whether the
whole or part of the material is concerned, specifically the rights of translation, reprint-
ing, reuse of illustrations, recitation, broadcasting, reproduction on microfilms or in any
other physical way, and transmission or information storage and retrieval, electronic
adaptation, computer software, or by similar or dissimilar methodology now known or
hereafter developed. Exempted from this legal reservation are brief excerpts in connection with reviews or scholarly analysis or material supplied specifically for the purpose
of being entered and executed on a computer system, for exclusive use by the purchaser
of the work. Duplication of this publication or parts thereof is permitted only under the
provisions of the Copyright Law of the Publisher's location, in its current version, and
permission for use must always be obtained from Springer. Permissions for use may be
obtained through RightsLink at the Copyright Clearance Center. Violations are liable to
prosecution under the respective Copyright Law.
The use of general descriptive names, registered names, trademarks, service marks, etc.
in this publication does not imply, even in the absence of a specific statement, that such
names are exempt from the relevant protective laws and regulations and therefore free
for general use.

Printed on acid-free paper

Springer is part of Springer Science+Business Media (www.springer.com)

For Meg

About the Author

David A.J. Seargent holds an M.A. and Ph.D., both in Philosophy, from the University of Newcastle NSW, where he formerly worked as a tutor in Philosophy for the Department of Community Programs/Workers' Educational Association external education program. He is also an avid astronomer and is known for his observations of comets, one of which he discovered in 1978. Together with his wife Meg, David lives at The Entrance, north of Sydney on the Central Coast of New South Wales, Australia. He is the author of three published astronomy books; "Comets – Vagabonds of Space" (Doubleday, 1982), "The Greatest Comets in History – Broom Stars and Celestial Scimitars" (Springer, 2008), and "Weird Astronomy – Tales of the Unusual, Bizarre, and Other Hard to Explain Observations" (Springer, 2010). Currently he is the author of a regular column in *Australian Sky & Telescope* magazine.

Preface

It is said that Baden Powell advised anyone hiding from pursuers to climb a tree because people rarely look up! With due respect to tree-climbing escapees, this is a great pity. So many wonders fill our skies. From the far reaches of the cosmos to the air flowing around our heads, the sky is filled with beauty and mystery.

Among the Ancients, there existed a cosmology which saw the Universe as divided into three heavens. The First Heaven was described as being where the birds fly or, in our terminology, the atmosphere. The Second was the Heaven of the stars, i.e. outer space or the astronomical universe, and the Third Heaven was the home of the Creator. The First and Second heavens of this old cosmology together make up the sky. All atmospheric phenomena, all astronomical events, have their homes there. Some of the spectacles of the sky are well known, others are rare and yet others are truly mysterious. During the course of this book, we will meet members of each class.

In this book, we will look chiefly at events within the atmosphere – "weather" in the very broadest sense. Yet the gaseous envelope that embraces our world is not an isolated system. Our planet, including its atmosphere, is a part of a far wider environment. There is no clear dividing line between air and space, and nearly all meteorological phenomena have their deep roots in the astronomical. At the most basic level, there would not even be an atmosphere without the Sun's heat!

Our story therefore begins with the Earth in its cosmic setting; a lonely blue globe in the vastness of space. The scene looks a hostile one at first glance. Our world appears terribly vulnerable. Yet, as we look more deeply, we see how a wonderful set of circumstances conspires to make our planet the ideal home for a species such as us. In many respects, our habitable world may be regarded as the greatest wonder of air and sky!

But if the habitability of Earth depends upon such a fine balance, is it implied that this is readily upset and catastrophe easily precipitated? Can the actions of humanity, or some random natural event, bring about our end? Many prophets of doom tell us that our actions are already doing just that. But are they correct? What does the history of the interaction of Earth, atmosphere and space tell us about long-term changes in climate and habitability?

From these background considerations, we move to the denizens of the sky itself. First of all, we look at the spectacular light displays occurring at the very interface of atmosphere and outer space. The study of aurora or "polar lights" is one place where meteorology and astronomy meet, as a true understanding of this beautiful phenomenon cannot be understood without encompassing both disciplines. But auroras are not confined to Earth. As we shall see, even grander displays occur elsewhere in our solar system.

As we move deeper down into our envelope of air, we encounter other spectacles such as lightning, mirages, vortices and a wide range of other atmospheric phenomena. We look at these, not just from the perspective of our own planet, but from that of other Solar System members as well.

During our journey, we will encounter tales of some very controversial observations and phenomena that simply don't appear to fit into accepted knowledge. What are we to make of the various types of luminous objects that have been reported from time to time, either high in the air or close to the ground? Or mystery sounds? Or meteor-like events that seem to occur at much lower altitudes than "genuine" meteors? Then there are clouds that rumble, auroras and lightning strokes that "hiss" and sparks that jump out of patches of fog. And what can one say about something that looks like a meteor appearing *within a darkened room*!?

But merely reading about the sky with its odd and interesting denizens is no substitute for going out and observing for oneself, and to this end, a number of observational exercises are included within these pages. These range from simple observations of the Moon illusion, through seeking out Haidinger's brush in a polarized twilight sky, to estimating atmospheric opacity from monitoring the darkness of lunar eclipses, to looking for unusual lightning or controversial "sleeks" (the name given to those alleged meteor-like phenomena apparently occurring at low altitudes). Some of these make for entertaining demonstrations at star parties (exposing the Moon illusion is a case in point) while others raise the possibility of making real discoveries (a once-and-for-all demonstration of either the illusory or non-illusory nature of "sleeks" by a group of experienced meteor observers would be nice!). But most importantly, they bring us face to face with that wide open space on whose shore we all make our home; the vast and magnificent realm of air and sky, the realm of weather and all manner of wonders.

Because their hobby involves being out of doors at night, amateur astronomers are amongst those who have the greatest familiarity with the sky and what happens there. Clouds may not be their best friends, but astronomers probably notice them more than most! And just by being outside during the night hours, and aware of the phenomena around them, astronomers are more likely to encounter – and to notice – events (even those not especially associated with the sky per se) to a degree that many non-astronomers may not. Their familiarity with the sky and its denizens will also enable them to more readily recognize what is and what is not "normal" in the heavens. So even if the subjects covered in these pages may not always be strictly astronomical, it is hoped that amateur astronomers, especially, will be stimulated by this journey through the realm of sky and night and come to an even greater appreciation of this wonderful (and sometimes weird!) world in which we live.

NSW, Australia   David A.J. Seargent

Acknowledgements

This book is the product of long-time interests in astronomical and meteorological phenomena, including reports in both fields that do not quite fit into accepted knowledge. So many people have contributed to these interests over so many years that even attempting to trace them all would be impossible, so may I simply offer a profound "thank you" to everyone who has aroused my curiosity about these topics.

More specifically, I extend my thanks to Mr. John Watson and Ms. Maury Solomon of Springer Publishing for their encouragement and advice and for their faith in a volume which some might consider a little "left field" for the Astronomers' Universe series.

My hope is that you, the reader, will have your curiosity aroused by this tour through the wonderful, sometimes spectacular and at time mysterious denizens of our strange skies.

David A.J. Seargent
The Entrance

Contents

Contents

1. This Island Earth

A Lonely Blue Globe

For those of us old enough (sorry, mature enough) to remember the Apollo Moon landings, what is the image that first springs to mind when those heady days are recalled to memory?

There are certainly many choices. The stark lunar landscape of "magnificent desolation" as Buzz Aldrin described it, the raising of the US flag as the first men from our planet set foot on another world, or maybe the dawning of Christmas on the Moon beneath the orbiting *Apollo 8*, as the astronauts each read sections from the opening chapter of the Book of Genesis. These are all moving memories, but perhaps the most iconic of them all is the scene of distant Earth rising above the lunar landscape. The contrast said it all; a blue and white globe amid the stark blackness of space poised above a harshly beautiful, but totally hostile landscape. The contrast, the loneliness, the fragility of our home planet was so striking that it is not for nothing that the Apollo Moon program is credited by some as the catalyst for the age of environmental awareness on Earth. Mankind went to the Moon – and discovered Earth! Not the Earth that had hitherto been our whole world, strong and indestructible, but Earth the planet, isolated and alone; a tiny island of life in a vast ocean of sterility.

If space exploration has taught us one thing, it is the "anomalousness" of our home planet. In fact, the more we learn about the universe around us, the more anomalous Earth appears to be. Back in the days when little was known about the other worlds of the Solar System, these were pictured as being (more or less) other Earths. There was a wide assumption that they were populated. The great astronomer William Herschel, for instance, not only thought that the Moon was inhabited, but he also believed that beneath the hot surface of the Sun there was a cool world where intelligent creatures lived. What we see as the "surface" of the

D.A.J. Seargent, *Weird Weather: Tales of Astronomical and Atmospheric Anomalies*, Astronomers' Universe, DOI 10.1007/978-1-4614-3070-4_1, © Springer Science+Business Media New York 2012

Fig. 1.1. Earth rising over lunar horizon as photographed from Apollo 8 (Credit: NASA)

Sun, Herschel believed to be more of an atmospheric envelop shrouding the green and fertile world beneath!

Solar beings (never one of Herschel's more popular ideas) have long vanished, as have Moon men and Mercurians. Martians lingered longer, with some writers even proposing their existence as late as the 1960s.

Following the first probes to Mars and Venus in the early 1960s, the true nature of our planetary neighbors started to become clearer; and for those who cherished the notion of extraterrestrial counterparts of human beings, the emerging picture was not encouraging. Not that the absence of H. G. Wellsian Martians was of itself a revolutionary discovery. Except for a mere handful of extreme hopefuls such as the unconventional writers alluded to above, few people with knowledge of the solar planets still believed in them when *Mariner 4* reached Mars in 1965. What did take most people aback was the apparent hostility of the Martian environment to life of any sort. The first images from Mars actually made the place look less inviting than later research found it to be, but a Mars that did not obviously support an abundant growth of lichen-like vegetation so widely believed in prior to 1965 came to many as a blow against the very idea of abundant life in the cosmos as a whole. In short, if the planet deemed most similar to Earth did not even support lichen the confidence that there were human analogues populating worlds further afield suddenly looked less secure.

Planets, Planets Everywhere (But Not a Single One Like Home?)

With the discovery, from the 1990s, of planetary systems beyond our own, our understanding of Earth's place in the order of things took another turn, but not the one that many people had expected. Just as pre-space-age concepts of the other planets of our Solar System saw these as being (more or less) Earth-like, so pre-1990 ideas about other planetary systems pictured these as basically clones of the Sun's family. Not in detail of course, but the general "plan" for the majority of solar systems was thought to be one of small rocky planets orbiting relatively close to the parent star and giant gaseous orbs located at greater distances. Essentially, "Earths" and "Venuses" close in and "Jupiters" and "Saturns" further out.

The very first planets discovered outside the Solar System were so bizarre that they simply did not fit the mould at all. They were at the "Earth" end of the size spectrum, but orbiting a neutron star or pulsar – the ultra dense remnant left behind by a massive star blown asunder in a supernova explosion. Nobody had expected to find planets in such locations. It seemed incredible that planets could survive their parent star going supernova. Presumably, these objects were not original companions of the ill-fated star. They must have snowballed together from the debris left over from the explosion. The system was, clearly, an anomaly and it was quite pointless to compare it with the Solar System. But the first "regular" solar systems – systems of planets around more or less sun-like stars – provided an even bigger shock. These were the systems of hot Jupiters or gas giants wheeling around their stars in orbits whose radii made Mercury look distant!

The early technique of planet detection relied on the wobble of the parent star caused by the nearby planet's gravitational tug upon it. For this reason, hot Jupiters were the only ones capable of being found during the first several years of planet seeking. The planet needed to complete an orbit of the star for the detection to be made, so the technique greatly favored large planets having very short orbital periods and, *ipso facto*, very small distances from their parent star. But that did little to ease the shock. The very existence of such planets was the real surprise. Perhaps if the first hot Jupiter discovered had been a one-off like the pulsar planets, it could have been dismissed as an anomaly. But it was not a one-off. Hot Jupiters kept turning up again and again. Then, after enough years of data collection had elapsed to permit the detection of more "temperate" Jupiters (those orbiting further from their central stars and having longer orbital periods) another surprise was waiting. Unlike our own system's giant planet, most of *these* "temperate" Jupiters orbited their suns in highly eccentric orbits. Some of them moved more like comets than planets! A big surprise.

In the face of this, astronomers began asking themselves if our familiar Solar System was really the norm at all. Maybe these other planetary systems were the regular ones and *ours* was the anomaly!

We still do not know the answer to that question. Most of the other solar systems discovered to date have not been clones of ours, although the observational selection already mentioned partially explains this by making those systems harboring very large planets at small distances from their central stars the prime targets for discovery. Still, with longer periods of observation now enabling the detection of planets at distances from their stars comparable with that of Jupiter from the Sun, a few approximate solar system analogues have emerged. Yet, to date, nothing has been found that *exactly* replicates our home planetary family. Other methods of planetary sleuthing (detection of planets transiting their parent stars – evidenced by very slight drops in the light of the star – gravitational lensing and even direct observation in a few cases) have also added to the tally of extrasolar worlds and families of worlds and it may not be long before the question of just how typical the Solar System is can be answered with some degree of confidence.

Whether our Solar System is or is not more "typical" of its species than the often odd and sometimes outright bizarre systems discovered during the last couple of decades, one thing is sure. Not every planetary system is like our own. Astronomical discoveries during the latter years of the twentieth century and the opening years of the Twenty-First have taught us that there is a great variety amongst the solar systems out there. The lesson has a definite whiff of *déjà vu* about it. It is just another level of the lesson that, only a very few decades earlier, space probes taught us about planet Earth. Other planets are not simply different versions of Earth and, it is now becoming clear, not all solar system's are modeled on the Sun's.

For many people, the interesting question raised by all of this is "What about life beyond Earth?" "Is the prospect of finding life elsewhere altered in view of these discoveries?" Or, to approach this from a contra point of view "Is Earth especially fortunate in being the verdant planet that we know?"

The answers to these questions imply a more fundamental one. "Can life exist under a wide range of circumstances, or is it dependent upon a very narrow set of parameters?"

A lot of talk about the possibility of extraterrestrial life focuses on the relatively recent discovery of so-called extremophiles; living organisms that can exist and even flourish under conditions previously thought impossible. We have all heard account of the cores of nuclear reactors becoming choked up with the prolific growth of diatoms. These little creatures cannot just endure, but happily flourish under, conditions of radiation so intense that even large molecules are disrupted. Then there are the "water bears" or "moss piglets" (more formally known as *Tardigrades*); tiny but incredibly tough little critters that not only can survive the extremes of cold but are also capable of being desiccated, left apparently for dead, and then brought back to life and health by adding a few drops of water! But even these don't hold a candle to the biomass beneath our feet – *way* beneath our feet. Deep within the rocky mantle of this planet lives the remarkable *Bacillus Infernus*. Cut off from the rest of the terrestrial ecosystem, these micro-organisms would not miss a beat were the entire atmosphere of Earth to be blown off into outer space!

Fig. 1.2. Scanning electron micrograph of Tardigrade *Hypsibius dujardini* (Credit: Bob Goldstein)

The lesson to be learned from extremophiles seems to be that relatively simple organisms can live under a wide range of conditions. It is not impossible that organisms which we would think of as extremophiles might really live on Mars and/or Titan, Europa, in the higher atmosphere of Venus and maybe even in water-bearing asteroids and large comets. Interesting (if not always convincing) arguments have been put forward for life on each of these locations and there is even some purported evidence (albeit controversial) in support of these. The popular press, and even some scientists, have at times taken this a step further by suggesting that if life is ever confirmed in any of these locations, the existence of highly complex organisms of the type capable of human type minds is almost a foregone conclusion ... *somewhere*. Not, to be sure, on any of these solar system bodies, but on many other more Earthlike worlds orbiting the far flung stars of the Galaxy.

Let it be said straight away, that this is far from being a watertight argument. The very fact that we speak of *"extremophiles"* gives the show away. If these organisms could talk, they would not say that their environments were extreme. *We* think of them as extreme, only because we, along with most other more complex forms of life, could not survive where they thrive. But that is only to admit that we are far more fragile than they, with remarkably narrow zones of tolerance. If the extremophiles indeed could talk, one might ponder what they would call us!

The simple nature of the extremophiles does not support the assumption that life in extreme environments will eventually evolve into more complex organisms fitted to these environments. Life is certainly adaptable. There is no argument about that. The real question is "How adaptable?" The relative simplicity of extremophiles, plus the narrow range of biodiversity in extreme environments, clearly tells us that there is a limit to this adaptability. It also tells us that these limits are increasingly restricted the higher up the scale of biological complexity we go.

The really interesting issue is not to define the limits of life per se, but rather to try to define the limits of the sort of life that could possess human minds; that sub set of organisms which Poul Anderson called *androdes* (in no way to be confused with andr*oids*!) and which C. S. Lewis, in his space fiction trilogy, called *hnau.* These are beings with which we might meaningfully communicate, share information about astrophysical problems or discuss the meaning of life.

Anderson, by the way, used the term "androde" in favor of the more popular "humanoid." He did not equate the two, and actually avoided "humanoid" as much as possible. This latter term, he pointed out, refers primarily to the physical shape and form of an organism, but his preferred "androde" related to mental characteristics, something which Anderson found far more interesting. Humanoids are not necessarily androdes. For instance, modern humans would find little in common with *Australopithecus*, although I doubt if we would have much trouble terming this extinct creature a humanoid. Whether the reverse is true – whether androdes need not be humanoids – is a more intriguing issue. Science fiction writers have happily populated their imaginary worlds with non-humanoid androdes – the infamous "bug-eyed monsters" or BEMs – but sober biologists note that there are many advantages to mental development found in humanoid features such as bipedalism and the possession of suitable hands and arms. An androde need not look strictly human of course, but it may be true that most would, in Willy Ley's words, be mistaken for humans "by somebody who does not see very well and cannot find his glasses."

Nevertheless, the only example of such creatures of which we are aware is ... us! Other more or less intelligent species share this planet, but none would qualify as an androde or hnau. The appearance of just one species after more than four billion years on just one of the planets in the Solar System is, we must admit, not an encouraging sign!

In the past, astronomers have generally been quite upbeat about the possibility of finding androde life elsewhere, while biologists were much more sanguine. (Neither used this term, which appears to have been Anderson's own invention, but it is such a good one that we will employ it throughout the present discussion. Its usage avoids the ambiguity involved in "intelligent life." That expression might mean beings like humans or it might mean something like dogs and cats, both of which clearly qualify as being "intelligent," as anyone who has had pets well knows).

The reason why astronomers have widely believed in extraterrestrial androdes is twofold. First, the sheer number of stars in the universe strongly implies a vast number of planets and this in turn suggests a large number of environments suitable for life, including androde life. The discovery of other planetary systems has partially verified the first part of this, although as we have seen, results thus far have been less secure with respect to the second.

Secondly, astronomers deal with relatively simply objects operating according to basic physical laws whose operations are amenable to fairly straightforward mathematics. To be sure, the math can become involved and there will no doubt be many who take issue with the "relatively simple" judgment, but when compared to the story of life, the evolution of stars, galaxies, planets and whatever other astronomical object we might name, *is* "relatively" simple. The processes governing it appear to be basic and constant across the observable universe. Life, on the other hand, is the most complex phenomenon in the known universe. The number of variables is staggering and we are only just beginning to scratch the surface of how life interacts with and modifies its environment and how, in turn, its environment interacts with and modifies life. Much is still shrouded in mystery, especially when we start asking questions about the nature of mind and consciousness. An astronomer might feel happy with the broad picture of collapsing cosmic clouds breaking up into protostars, which in their turn collapse into stars sitting in the middle of rings of material that accretes into planetary systems, but when biologists look closely at the finer details of what happens on those planets and contemplates the great gulf between the most complex abiotic molecules that might form there and the simplest living organism, the simple "process A is followed by process B" approach begins to look far too trite. And the jumble of processes between the first living cell and human beings hasn't even started! One astronomer once poured scorn on skeptics as being "absurdly anthropocentric", and this in general is reflected in public opinion. But biologists are more likely to take a less dogmatic view.

In more recent years however, some astronomers have come over to the biologists' side. Since the turn of the present century, several books have appeared (some of them even written by astronomers!) giving openly skeptical assessments of the probability of widespread androde life – some even of the very existence of extraterrestrial androde life itself. Examples of such books include S. R. Taylor's *Destiny and Chance: Our Solar System and its Place in the Cosmos*, P. Ward & D. Brownlee *Rare Earth: Why Complex Life Is Uncommon in the Universe*, G. Gonzalez & S. Richard *The Privileged Planet* and W. C. Burger *Perfect Planet, Clever Species*.

Let it also be said that these books, together with other volumes and articles taking a similar line, have come in for some strong criticism. Yet, the very fact that they were written and published at all is a sign that the popular picture of a universe teeming with androde life no longer has a monopoly of thought, not even amongst astronomers.

Why the change?

In essence, it is because the factors making for an environment suitable for androde life are now seen as being more numerous and the situation more complicated than had earlier been imagined. Although popular media presentations often appear to suggest that finding other androde life is simply a matter of finding a planet more or less similar to our own in an orbit that is far enough from its star for water to exist as a liquid but not so far as to be permanently frozen, the full story is now looking a lot more complex. Let's have a look at some of these "complexities" and try to figure what they mean for life on Earth and what they might mean for similar life elsewhere.

The Effect of the Wider Universe

The universe itself appears to be finely balanced in more ways than one and cosmologists see this balance as quite literally a matter of life and death. Or, to be more precise (as the very word "death" implies that something has first been alive!) a matter of life and never-life. Of course, as this fine balance affects the universe as a whole, it has relevance for the biological potential of any planet and says nothing as to whether Earth is or is not especially suitable for life. Still, the facts are so remarkable that they are surely worth at least a brief mention.

Probably the most basic fine balance is the rate at which space is expanding, which in turn depends upon the amount of matter in the observable universe. These both seem to be at a critical level permitting the universe to avoid re-collapsing on the one hand and flying apart on the other. Too much matter (too slow an expansion) would cause the universe to stop expanding and go into a contraction phase followed by total collapse before stars and planets had time to provide an environment for life. Too little matter (too rapid expansion) would dilute the universe of matter before galaxies, stars and planets had a chance to form.

As it is, in the real universe, primordial gas (mostly hydrogen and helium) gather into vast clouds which by means of processes still not fully understood, form galaxies containing hundreds of millions of stars. But stars are not immortal. They eventually grow old and die and the largest and brightest are the ones that age faster and die youngest. By way of comparison, our Sun (a middleweight star) has been around for some five billion years and is roughly at the midpoint of its life expectancy, but the largest and hottest stars are lucky to make it to ten million years of age; mere infants-in-the-cradle. These large stars are also the ones that die in the most spectacular fashion, blazing into brilliant supernova which for a few months may outshine the combined light of all the other hundred thousand million stars of the home galaxy. But because stars are essentially thermonuclear furnaces in which primordial hydrogen and helium are converted into elements of greater atomic mass (somewhat confusingly called "metals" by astrophysicists), the death throes of these giants return to the interstellar medium within their galaxies, the material from which they initially formed, albeit processed and enriched with "metals." New generations of stars form, die, and enrich the medium still further, until there are enough heavy elements to form both planets and the materials from which living things are composed.

Stars like the Sun live far longer as we already mentioned, and die in a less dramatic manner. After about ten billion years, they swell out into red giants whose diameters are of the same order as those of the orbits of planets such as Earth and Mars, while their centers collapse into dense planet-sized hot cinders called white dwarfs. The outer layers are puffed away as "planetary nebulas" (so called because some of them show as small disks in telescopes which look a little like nebulous planets) which gradually disperse into interstellar space and lose their identity, leaving behind the white dwarf core.

The majority of stars – smaller, fainter and cooler than the Sun – live extremely long and sedate lives and fade away with neither supernovae nor planetary nebulas to mark their demise. However, like the proverbial candle that burns brightest just before it goes out, these dwarf stars will have a period of glory just prior to their final fade as they brighten into Sun-like splendor. I say "will", because the life span of these stars is so long that the universe itself has not yet aged sufficiently for any to have reached this brighter terminal phase. Only astrophysical theory predicts that it will happen. Because of their long life and sedate lifestyle, stars such as these (and they form the majority) do not recycle processed material back into the interstellar medium.

A universe in which slow initial expansion turned to contraction and collapse after only the first couple of generations of stars formed would not be one in which life could be present, for the simple reason that there would be insufficient "metals" from which planets and the living organisms that might inhabit them could be composed.

But the expansion of the universe is not the only, nor even the most stringent, example of universal fine tuning. As the brilliant and controversial British cosmologist Sir Fred Hoyle pointed out, if the stable resonance level of Carbon 12 differed ever so slightly from its real value, carbon based life would be impossible anywhere in the universe. Moreover, that property of carbon is itself dependent upon the exact strength of the forces within the atomic nucleus, so if these had differed ever so slightly from their actual magnitude, nobody would be around to know.

Similarly, if the force of gravity was stronger, stars would become more compressed. This in turn would raise their internal temperatures, accelerating the thermonuclear reactions within their cores. In consequence, their life span would be shorter and there would be no time for the formation of planets suitable for complex life. On the other hand, weaker gravity would mean cooler stars with little or no nuclear fusion to synthesize heavy elements.

These are not the only examples of universal fine tuning. Far from it. But it is enough to show that the universe is really a "Goldilocks" cosmos from the point of view of life.

Some scientists suspect that the values of physical (apparent) constants such as the relative strengths of the "strong" and "weak" nuclear forces change over the extent of the universe at large and that the "Goldilocks" situation may not be truly universal. The more usual assumption however, is that these constants are truly constant and that our region of the universe is representative of the whole.

Goldilocks Strikes Again!

As we turn from the universe at large however, and focus down onto smaller and smaller details of our environment – our galaxy, our region of that galaxy, our Solar System and, finally, our planet – does the assumption that the Goldilocks nature of that environment is representative of the whole hold up or does it become increasingly tenuous?

Imagine that we are aliens from some other dimension, or whatever, in search of suitable places in this universe where we might find forms of life capable of communicating with us. We have already determined that the universe as a whole has the right parameters to permit the existence of such life somewhere within it, and it is now our task to search out those places where more "local" conditions allow this potential to bear fruit. We want to look for suitable galaxies, suitable regions within those galaxies, suitable stars within those regions and suitable planets orbiting those stars. The question we want to answer is whether suitable galaxies, regions of galaxies, stars and planets are the rule or the exception. Will our search turn up vast numbers or just a few? Will each step of our search be a journey along the highway of normality, or will it take us ever further into the byways of the anomalous?

A Goldilocks Galaxy?

We begin our search by honing in on the Milky Way, home galaxy of the human race. The Milky Way is a giant spiral galaxy; one of a pair that dominant a small group of galaxies known as the Local Group.

Fig. 1.3. The "Pinwheel Galaxy" M101. Located 25 million light years away, this galaxy strongly resembles the Milky Way. Seen face on, our own galaxy would look very like this image. This image is a composite of 51 individual Hubble exposures in addition to elements from ground-based photographs (Credits: NASA and ESA, J,-C, Cuillandre/Coelum, G. Jacoby, B. Bohannan, M. Hanna/NOAO/AURA/NSF)

The other member of the pair is the Andromeda Galaxy (M31) and, as it is some two billion light years away, we see that the term "Local" is being used here in a way that only astronomers could justify!

Now, spiral galaxies are not the most common types. Indeed, they constitute only about 5% of the galactic menagerie yet they appear to be the ones most likely to harbor life bearing planets.

Why is this?

The reason goes back to the evolutionary process mentioned above where large stars of one generation enrich the interstellar medium with heavy elements from which a more "metal" rich succeeding generation is born. It seems intuitively obvious that stars forming from material that has been significantly enriched with heavy elements are the ones that we would expect to have planets in orbit around them and, indeed, this has been precisely what has been found. The Sun is metal rich and so are the other stars for which planets have been confirmed. In fact, many are more metal rich than the Sun and it may well be that the stars which possess very large planets (which, in the main, are also the ones most likely to be found using present day techniques) are the ones that have a larger store of metals than our own Sun. That too seems to make good sense.

But not all galaxies are suitable for the progressive building up of metal rich interstellar media and the continuing chemical evolution of stars. Small irregular galaxies are very poor in metals, presumably because they were late starters in the star forming business. But very large "elliptical" or "globular" galaxies don't fare that much better, although the reason here is quite different. These systems are mostly comprised of small red stars of the type that are not really into recycling. Larger stars presumably once blazed there as well, but it seems that something removed the interstellar medium from which further generations of stars could form and once these larger and shorter lived stars had their day, no new generation succeeded them. Only their smaller and very long lived siblings remained to light the galaxy.

Moreover, deep within the central cores of these systems, giant black holes lurk and the density of stars is high. The orbits of the stellar members of such galaxies are ellipses that carry them into these inner city regions of close stellar encounters and energetic bursts of radiation from stars being tidally shredded and pulled into the central black hole. Even if planets did exist around many of the stars in such galaxies, this situation does not sound a very inviting one for life.

It is now thought that most of these giant globular systems were formed by the collision and merging of two or more separate galaxies. Indeed, around three billion years from now (don't panic! I said "billion" not "million") the Milky Way and M31 will merge into just such a supergiant globular. It is very unlikely that any stars collide in such mergers, but the interstellar medium is rammed right out of the galaxy. In clusters richer than the Local Group, there is a greater chance that the largest galaxies merged eons ago, which probably has a lot to say about why our own galaxy and its rich interstellar medium occupy only a puny cluster.

To have accumulated the requisite store of metals however, a spiral galaxy presumably needs to be large, which is also confirmed by the fact that we *do* live in such a galaxy. A large galaxy in a small cluster of galaxies: Could that be the galactic (and intergalactic!) portion of the formula for a suitable abode of androde life?

On the reverse side of the coin, a galaxy experiencing a high rate of star birth might have its own problems. The birth of many stars implies a relatively high rate of supernova when the more massive examples die. This, in turn, implies that more regions of the galaxy are in danger of experiencing bursts of deadly radiation than is the case in a quieter galaxy. Furthermore, it might also be suggested that chemical evolution in such a system could proceed a little too quickly; passing rapidly beyond the delicate chemical balance where planets having the happy balance of carbon and oxygen are likely to form. Recent observations courtesy of the citizen science *Galaxy Zoo 2* project have confirmed earlier hints that spiral galaxies with central bars (of which class the Milky Way is a member) have slower star formation rates than galaxies that are otherwise similar, save for a bar. This raises an interesting question. Had the Milky Way been a "normal" rather than a barred spiral, would we be here today?

Another feature of our home galaxy might also need to be factored in at this point. Recent surveys of the realm of galaxies have pinpointed a very unusual feature of the Milky Way. It has two close companions which are not only larger than run-of-the-mill dwarf companion galaxies, but are also (and very unusually) in a state of vigorous star formation. The Magellanic Clouds/Milky Way system is unlike anything else that has as yet been found in galactic surveys!

As we will see during the course of this chapter, anomalies seem to play an important role in determining suitability for androde habitability. It is as if habitability runs against the much cherished *Assumption of Mediocrity* i.e. the assumption that our location in the universe is not special in any fundamental sense. Of course "special" is a loaded word, but if it is replaced by "non-anomalous" we will increasingly see that the opposite is actually the case. Or, at least, that is how it appears. Just count the anomalies upon which Earth's capacity for androde habitability depends as you read through this chapter!

This is jumping ahead of ourselves, but the point is raised here simply because the Magellanic Clouds provide another anomaly and we might ask whether this can also be explained *anthropically*, i.e. Are we seeing this anomaly simply because we would not be around to see anything at all if the anomaly did not exist? Or, in other words, "Does this anomalous Magellanic Clouds/ Milky Way grouping make our galaxy more friendly toward androde life?"

On the face of it, it would seem hard to believe that it does and we are tempted to brush this aside as a mere coincidence. True coincidences must happen sometimes and we might be too over-zealous to see anthropic hints in all of them. Yet, in saying this, it is interesting to note that a flow of hydrogen gas known as the *Magellanic Stream* links the Clouds with the Milky Way and is thought to have provided our galaxy with fresh star forming material for the past billion years. Of course, that is too recent to have had any effect on the Solar System (which formed about 4.5 billion years ago), but it is not impossible that an earlier steam may have pumped up the rate of star formation in the youth of our galaxy. The more stars, the faster the accumulation of heavy elements suitable for planetary formation. From an estimate of the amount of "metals" in the galaxy today, it does indeed appear that the supernova rate was far higher in the early years, presumably indicative of higher star formation rates then, though whether this relates in any way to an early Magellanic Stream remains pure speculation. (It should also be mentioned that it is a good thing for us that the supernova rate *has* declined over the eons. We don't want too many of these cosmic bombs going off in our neighborhood today!)

Probably the most common type of galaxy is the dwarf elliptical. Very different from the Milky Way, the Magellanic Clouds or any of the galaxy types about which we have just been speaking, these faint assemblages of small stars are predictably metal poor (although some have a higher content of heavy elements than others) and do not stand out as likely homes for planets capable of supporting androde life.

Let us now hone in a little closer on our galaxy. We find a central bulge of old and mostly red stars surrounding a relatively small central black hole. That component of the galaxy is like a scaled down version of a giant globular galaxy, with most of the negative attributes from the viewpoint of life-suitable worlds.

Surrounding this is a disk of stars and interstellar material. Actually, there are two components to this feature, the thick disk and the thin disk, but it is within the thin disk (actual thickness around 1,000 light years) that most of the interesting activity takes place. This is where generation after generation of stars are born and die and where a constant haze of interstellar material is constantly being recycled and enriched by precisely this process. Not surprisingly, it is within the thin disk that the Sun and its planetary entourage is found.

Yet, a little closer examination discloses a further complication. Not everywhere in the thin disk is equally suitable for androde life. Too close to the galactic center will mean exposure to the overcrowding and violence of inner-city galactic life. The central black hole may not be as massive as those in the giant elliptical or globular galaxies, but it is still large enough to be dangerous. Moreover, the heightened possibilities of nearby supernova explosions bathing the Sun's in lethal radiation and of close encounters with other stars or molecular clouds gravitationally disrupting the cloud of comets that surrounds the Solar System and sending showers of them down amongst the planets, makes these inner regions dangerous places indeed.

Yet, too far out in the suburbs is not too healthy either. The problem with the outer regions of the disk lies with the relative paucity of stars there. Fewer stars means slower recycling of processed material and this in turn means slower chemical evolution, reduced likelihood of planets and therefore of life. Between these two extremes, there is a "Goldilocks zone" in which – at about 28,000 light years from the galactic center – our Sun orbits. The width of the galactic habitable zone is unclear, but the fact that the Sun is about half way between the galactic heartland and the outer fringes places it far enough away from the central galactic hub to be clear of the violent events taking place there whilst keeping it within a region sufficiently populated to have built up a good store of heavy elements.

But the Sun's location in the galactic habitable zone per se is not the only factor in its favor. It also remains unusually close to the galactic plane for a middle aged star. Most stars of its age have had their paths around the galaxy pumped up by the gravitational perturbations of other stars and clouds of interstellar material until their orbits take them from 600 to 1,000 light years above or below the plane of the galaxy. The Sun's orbital oscillations are more typical of those for objects as young as one thousandths of its age, not venturing more than 250 light years from the central plane. This constant proximity to the plane undoubtedly helps in shielding the Solar System from too much radiation from the galactic core and also probably indicates that our system has experienced unusually few close encounters with stars and giant molecular clouds throughout its history.

Taking up this latter point, it does indeed appear that the Sun has been cruising through an anomalously quiet region of space for the past several million years at least. We seem to be in a part of the galaxy unusually free of interstellar material. Although the entire history of the Solar System may not have been as peaceful, there is reason to think that it has been more benign than that experienced by most stars, thanks to a rare synchronicity between the orbital velocity of the Sun around the galactic center and the speed of propagation of the galactic spiral arms. Let's look more closely at this.

Examining a photograph of a face-on spiral galaxy, the most obvious features are the beautiful swirling spirals of stars that give these systems their name (see Fig. 1.3 for a fine example). At first glance, it would appear that these spiral arms are unusually packed with stars, but in actual fact the star density within them is only some 5% higher than in the rest of the disk. What makes them so bright and conspicuous is the compression of interstellar material triggering vast stellar nurseries replete with large numbers of very bright blue stars; the type that live fast and die young. These are rare outside of the spiral arms, because they blow up before leaving their nurseries. They are gone while still within the spiral arm, but that is enough to make these arms as deadly as they are beautiful. If the Solar System were to pass through a spiral arm, chances are high that one of the stellar spendthrifts will blow itself to smithereens while it is close at hand, with potentially devastating consequences for life on Earth. But even if a supernova doesn't get us, there is a good chance that a shower of comets will. Tidal perturbations from the star-forming clouds strung out along the spiral arms send myriads of these objects in from the circumsolar comet cloud, posing more risks to terrestrial life. All in all, spiral arms are good places to keep away from!

And it appears that the Sun does keep away from them; more than most other stars at least. The arms themselves are not really material features "orbiting" the galactic core. Rather, they are density waves that propagate around the galactic disk. As the density wave passes, interstellar material is compressed into star-forming regions and the resulting crest of star birth is what we see as the spiral arm. Near the galactic center, the orbital velocity of

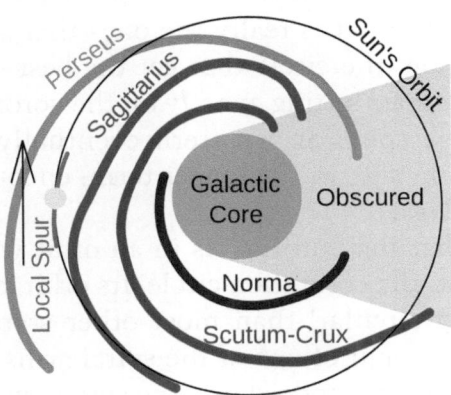

Fig. 1.4. Diagram of Milky Way Galaxy showing Sun's orbit (following J. H. Taylor & J. M. Cordes, 1993, *Atron. Jnl.* 411: 674–684)

the stars within the disc is larger than the speed of the wave. In the outer regions the speed of the wave is larger. Stars whose orbits are confined to these zones either catch up with the wave and pass through it on a periodic basis or else have it periodically sweep over them. Either way, they regularly encounter the dangerous spiral arms.

But it follows that if stars deep within the galaxy orbit faster than the spiral arms progress and those in the outer reaches go around the galaxy more slowly, somewhere in between these two extremes there must be a narrow region where the orbital velocity of the stars and the velocity of progression of the wave exactly match. There will be a sort of interface or line of demarcation between orbital velocities faster than the spiral arm progression and those slower than it, and it should be possible for a star at that position, yet located between the spiral arms themselves, to orbit the galaxy time and again without directly encountering the spiral arms. The experience of such a star reminds the writer of a report of a ship sailing blissfully along in the eye of a hurricane, enjoying calm and clear weather while all around the storm raged furious! While ever the speed of the ship matched the onward velocity of the storm, it remained safe in the region of calm. Stars like these are, in effect, doing the same thing on a galactic scale.

The line of demarcation has a special name. It is called the *corotation circle* of the galaxy. Its distance from the galactic center is known as the *corotation radius*. Although we have been speaking as if a star precisely on the corotation circle is the one that lives the charmed life, it is really the ones that are very close to, but not actually on, the circle that get the best deal. It has been shown that those stars sitting *exactly* on the corotation circle will resonate with the spiral arm pattern eventually getting ejected from their position and, in the longer term, encountering the spiral arms themselves.

It appears that the Sun travels in an orbit which is close to, but not exactly at, the corotation circle. Its orbit is also less eccentric (more nearly circular) than most other stars of its age and located on a minor spur of one of the spiral arms, in between the major arm patterns. It is indeed like the ship navigating the eye of a hurricane, keeping pace with the calm within the storm.

Recently, computer simulations by Robert Grand of the *Mullard Space Science Laboratory of the University College, London* indicate that being caught in a spiral arm may be even worse than originally imagined. Grand's work suggests that, rather than passing through the spiral arm relatively quickly, once a star enters a galactic arm it gets swept along by it and effectively becomes stuck there for around 80–100 million years, greatly increasing the time it is exposed to the dangers of the spiral arm environment. If this work is confirmed, the importance of the Sun's location in the "eye of the hurricane" looks even more important.

Nevertheless, the Sun's happy circumstance does not imply that the corotation circles of all spiral galaxies are situated in regions suitable for androde life. The size of the corotation radius depends on several factors, such as the size of the galaxy and its distribution of matter (which affect the orbital velocity of its constituent stars), the rate at which orbital velocities decrease across the diameter of the galactic disk and the speed at which the spiral waves themselves progress. In some galaxies, the corotation circle may lie dangerously close to the galactic heartland. In others, it may be way out in the metal-poor fringes. In some, the theoretical corotation circle may even lie beyond the confines of the galaxy altogether, so that *no* star lies close to it. For our home galaxy however, the corotation circle lies at a nice moderate distance – Goldilocks once again; not too close in, not too far out, but just right!

So the location of the Sun appears to be just right and it further appears that this "just right" position is far from typical of the galaxy's stars in general.

A Goldilocks Star?

But what about the Sun itself? Is it typical of the majority of stars in the galaxy, or does Goldilocks need to test its "just-rightness" also?

Apparently, it is also in need of Goldilocks' seal of approval. Although we tend to think of the Sun as an average star, the only thing "average" about it is its location on the spectrum of star sizes; just about mid way between the largest and hottest stars and the smallest, coolest and dimmest ones. Now, in most populations of objects having a range of sizes, the greatest portion of the population will consist of the smallest objects – the lower end of the range. It is this way with stars. Most of them are red dwarfs, far and away smaller and fainter than the Sun. If the Sun appears as a firefly when compared to the largest and brightest of stars, it is a searchlight in comparison to the red dwarfs, and it is these stars that form the bulk of the stellar population. Far from being typical or average, the Sun turns out to be in the brightest 5% of stars in the galaxy. It belongs to the G class or moderately hot stars, located about mid way on the spectrum of mature stars known as the Main Sequence. Stars on this sequence have reached the stable hydrogen fusion stage in which they spend the greater parts of their lives. Sometimes these stars are known as "dwarfs", even though that term is hardly fitting for anything much above the lower red end of the Sequence. The term is not really employed to suggest that Main Sequence stars are small per se, but rather to distinguish those at this stage of stellar evolution from the bloated "giant" phase which any star not at the fainter tail of the Main Sequence will enter upon reaching old age.

Main sequence stars of the Sun's type are comparatively stable objects that do not show large variations in brightness and heat output. This is very fortunate for us on planet Earth for reasons that are too obvious to need mentioning. However, from observations of other stars of solar type, some astronomers have concluded that the Sun is even more stable than most of its brethren. This conclusion has been challenged by others who suggest that an observational selection effect might be responsible for the relative quietness of the Sun. That is to say, because Earth orbits in the equatorial plane of the Sun, the measurements we take of it are necessarily "equatorial" ones and these zones are somewhat quieter than its polar regions. But other stars should be randomly orientated with respect to Earth and therefore any representative sample will be a mixture of both polar and equatorial orientations. The average level of activity will in consequence appear to be larger than that of the Sun. Numerical simulations testing this hypothesis have found, however, that it does not completely explain the difference. For some reason, the Sun does appear to be an abnormally quiet and steady star.

The composition of our Sun is also a little odd. As we said earlier, the chemical evolution of a galaxy results in its stars becoming richer in "metals" (in the eccentric astronomical sense of that word!) as time goes on. We also saw that this does not proceed steadily throughout the galaxy; dense inner regions evolve faster, in this sense, than the sparsely populated outer fringes. The Sun's region is about mid way between the extremes, so we might expect our star to be not especially rich in metals, whilst still having a larger component than similar stars out on the galactic fringes. Broadly speaking, this is true, but the Sun is rather richer in metals than we might have expected for a star of its location *and* age. It has about the same metal content as similar stars forming *now*, not the same as its five billion year old contemporaries. Moreover, the mixture of heavy elements also appears to be in "just right" portions thanks – apparently – to the contributions of at least four older stars from whose debris the Sun and its planets formed. From the analysis of minute extra-solar grains found in meteorites, astronomer A. Delsemme concluded that the material from which the Sun formed had been enriched by debris from one carbon rich star, another oxygen rich one, a third that was generously endowed with both magnesium and silicon and a fourth that was rich in iron! A different mixture or different proportions of these elements could have seriously disrupted the chemistry of life and prevented the existence of complex organisms on any of the Sun's family of planets.

On the other hand, most of the extrasolar planets discovered to date orbit stars that have a higher metal content than our Sun. So, although the Sun is unusually rich in metals for a star of its age and location, it is at the metal-poorer end of the spectrum for stars known to harbor planetary systems. Part of this could be a selection effect, but if it is, it is an interesting one in its own right. Planetary systems having very massive worlds, especially those in small star-hugging orbits, are the easiest to discover and it seems reasonable to expect that where heavy elements ("metals") are very abundant, the largest planets are likely to form. Moreover, these systems should also contain the greatest mass of rocky material and it is thought that the gravitational interaction between accreting planets and this "debris" is the prime cause of planetary migration, such that many of these very large planets will migrate inward from their cool birth places into the torrid, close in, orbits where they are found today. This may be the reason why such systems are prone to be found accompanying very metal-rich stars. Moreover, some of these stars may actually have swallowed one or more Jupiter-sized planets, and any amount of smaller asteroidal objects, also adding to the metal content of their outer layers, if not of their deeper regions. This leads on to another Goldilocks consideration, which we will come to in a minute, but first we should note yet one more solar oddity.

That is to say, as well as being a little bit odd in terms of its metal content, the Sun is also mildly unusual by being single. A recent survey of sunlike stars found that more than 60% were double or multiple, whereas an earlier survey found that single stars of similar type to the Sun constituted only about 30% of that class. The two surveys came up with much the same result; always an encouraging sign! Multiple stars are not necessarily without planets, but the chances are that there is less "left over" material in these systems with which to form substantial planetary systems and, for those planets that might form, the chance of stable orbits where precisely the right amount of heat and light is received for an environment suitable for androde life is not high.

But back to the further Goldilocks consideration noted above, it seems that there is yet another item in the galactic household that the young heroine of the Three Bears nursery story must test; the Solar System itself. It is of no use having a star that is just right pursuing an orbit that is just right in a galaxy that is just right. Solar systems come in such a variety of forms that it looks a safe bet that not just any old planetary system will do. This indeed appears to be the case.

Solar systems having large planets orbiting close to the central star – the so-called "hot Jupiters" and "hot Saturns" – are generally thought unlikely places for terrestrial worlds. The reason is simple. Gas giant planets such as Jupiter, Saturn and their extrasolar counterparts form in the cool outer regions of nascent Solar Systems, not in the regions close to the central star. To end up in these small orbits, the planet must migrate inward, meaning that at some time during its journey it must have passed through the regions where Earth like rocky worlds form. Now, the last thing that such worlds need is a migrating Jupiter! Earth like planets in these systems are either flung from the system altogether, ending up as orphan worlds wandering through the dark of interstellar space or they are hurled into the central star and destroyed. Either way, their prospects for evolving into homes of life is discouraging, to say the least!

Some astronomers are not quite so pessimistic about the possibility of hot-Jupiter systems harboring potentially habitable planets. They argue that while it is true that a migrating Jupiter-like gas giant will disrupt the orbit of any pre-existing inner planet, it is possible that in some systems at least, there may be enough remnant material *after* the gas giant has finished its inward journey to accrete into one or more terrestrial worlds. This suggestion might be within reach of observational testing before much longer, thanks to the advent of space based observatories dedicated to searching for small extrasolar worlds.

Whether or not some of the systems containing hot Jupiters can also harbor Earth like worlds, another class of planetary systems discovered in the past couple of decades almost certainly excludes this possibility. These are the systems in which a Jupiter-like planet moves in what by the standards of our own Solar System would be termed a cometary orbit. The very thought of Jupiter moving through the Solar System on the orbit of Halley's Comet is just too depressing to contemplate! Planets would be thrown out left, right and center!

The possibility of an Earth in a stable orbit in these systems is essentially out of the question. At least, for an orbit anywhere near the habitable zone. It is possible that small planets could orbit far beyond the eccentric gas giants (perhaps planets that had been hurled out of the inner zone, but managed not quite to escape the system altogether), but these would be of no interest for those seeking habitable worlds.

At the other end of the scale lie solar systems that have no gas giants at all. These are surely out there and the new generation of planet finders like *Kepler* might soon uncover them. Traditionally, these have been considered not very suitable places for androde life. This is because Jupiter was seen as acting like a shield to incoming comets that might otherwise collide with Earth. It was figured that in a system that did not sport a relatively large gas giant planet in an orbit roughly corresponding to the location of

Jupiter in our own Solar System, any potentially Earth like planet would be so heavily bombarded with comets coming in from the remote regions of the planetary system that any ecosystem would be constantly traumatized by massive extinction events. No sooner would the web of life start to recover from the last hit when another colliding comet would come along and wreck any progress toward recovery that had been made.

Yet, a Jupiter like world could not orbit too close in nor could it be too massive. The failure of a rocky planet to form in the asteroid zone of our home system, and probably the relatively small size of Mars as well, bear testimony to the disruptive effect that a gas giant can have. Fortunately for Earth, Jupiter is small enough and sufficiently far away to have allowed our home world to build to a respectable mass. A larger or a closer Jupiter might not have been so benign. Even if Jupiter had been larger, but not so large as to prevent Earth from forming, the results for our planet may not have been healthy. It seems reasonable to expect that a larger Jupiter would have made Earth's orbit more eccentric than it is now, with unfortunate consequences for climatic variations.

This line of reasoning led many astronomers to conclude that a moderate-sized Jupiter was necessary for Earth to turn out the way it has. If Jupiter had been absent or small, there would have been insufficient protection from incoming comets, but if it had been too large, there would have been too much perturbation of our orbit or, in the more extreme case, no single planet Earth at all.

While much of this continues to hold true, a study by J. Horner and B. W. Jones has cast considerable doubt upon the efficiency of Jupiter as a shield for the inner Solar System. They point out that when this idea was first floated, major Earth impacts were thought to come from long-period comets, i.e. those originating well beyond the orbit of the outer planets. More recent findings tell a different story however. Concerted efforts during the past couple of decades in search of potentially threatening objects have found that the danger from Earth-approaching asteroids and comets of very short period far outweighs that from long-period comets. Jupiter certainly acts as a shield against the latter. Indeed, the gravitational perturbations of the growing gas giant cleared the inner system of billions of these bodies. But it is not a shield against the currently more dangerous asteroids. In fact the rather surprising conclusion reached by Horner and Jones is that the presence of Jupiter actually *increases* the impact rate of asteroids and short-period comets. Earth would experience fewer impacts from these bodies if Jupiter was absent or very small on the one hand or if it was twice its real mass on the other. Yet, if Jupiter had only grown to around the mass of Saturn, the situation would have been much worse, with an impact rate on Earth about twice that actually experienced. These authors suggest that, far from being the "friend" that Jupiter has come to be portrayed, it is in reality a "foe", though not so great a foe as Saturn would have been, were it the Solar System's most massive planet and located in Jupiter's orbit.

Calling Jupiter a "foe" might be being a bit too hard on the old fellow however. Jupiter may indeed increase the rate of impacts, but that is probably not altogether a bad thing. The types of impacts involved are mostly low velocity ones unlike the average impact of long-period comets, whose highly eccentric orbits and normally high – even retrograde – inclinations make them potentially far more formidable missiles. Many scientists now suspect that impacts from asteroidal bodies, or fragments thereof, were responsible for Earth's store of organic compounds and water. If true, a moderate impact rate may have helped rather than hindered life. More dramatically, violent impacts large enough to trigger mass extinctions may have been "friends" to life in general (even though "foes" to the extinguished species!) by preparing the way for new species to fill the niches left by those that disappeared. Though disastrous in the short term they effectively pruned the tree of life, setting the stage for new growth to appear. Our presence here today, for example, in part depended on the dinosaurs becoming extinct.

A Goldilocks Planet?

Coming down to even finer Goldilocks specifics, we may ask as to *how* "just right" a planet needs to be for androde life. Or, in different words, how close to the Earth that we know did our planet need to be to become home for the human race?

We hear a lot about the habitable zone of planetary systems; the relatively narrow region where planets like Earth can support bodies of liquid water on their surfaces. Earth orbits at an average distance of 93 million miles or about 150 million kilometers from the Sun. For the sake of convenience and to equip ourselves with a handy measuring stick for solar system-scale distances, this distance has been given a name; one Astronomical Unit (AU). We know that the amount of energy received by Earth from the Sun at 1 AU distance is suitable for androde life. We would not be here if it were not! But just how far can a planet like Earth stray inside or outside of that distance (or the equivalent distance from a star of different energy output than the Sun) before it loses its habitability? That is the question.

According to astronomer J. Kasting, Earth could remain habitable anywhere from 0.95 to 1.37 AU; not a very wide region to be sure. Yet, we must also remember that not every planet within this zone is automatically habitable. There are what might be called "intrinsic" and "extrinsic" factors governing the potential habitability of a planet, and both of these sets of features must come together. The location of a planet with respect to the habitable zone is an extrinsic factor, but if the planet is intrinsically uninhabitable, it could be right in the middle of the zone and still be as sterile as a surgeon's scalpel. Our own Moon is a good instance of this. Although it is located as safely within the habitable zone as Earth itself, that certainly does not make it a world suitable for life.

And then there is Venus!

In terms of mass, dimensions and bulk composition, Venus is closer to Earth than any of the Sun's other worlds. It is also closer in space, so the impression might be gained that we have a second Earth right on our doorstep, cosmically speaking. Yet, we all know differently. Conditions on Venus are about as far from those of Earth as we can imagine.

The impression is sometimes given that the reason why Venus turned out to be so hostile rests solely with its location outside of the planetary habitable zone. At around 0.72 AU from the Sun, it is too close and therefore too hot to sustain complex life. This is, of course, part of the answer, but it overlooks some very negative intrinsic properties of this second rock from the Sun. If, let's say, the positions of Venus and Earth were reversed – so that Venus occupied what is in fact Earth's orbit and Earth, Venus' – Earth would indeed be uninhabitable. Yet, Venus would not be like Earth. It would be almost as hostile to life as it is in the actual Solar System. In fact, where ever Venus was placed, it would remain a hostile world.

Why?

Principally, because it's thick crust makes it intrinsically unsuitable as a habitat for complex life. Earth's crust is unusually thin (we will see why a little later in this chapter) and is broken into several crustal plates enabling a process to take place on Earth that is found nowhere else in the Solar System – plate tectonics. As the planet's crustal plates slide, collide and slip under one another in this constant dynamic process, heat from within the Earth is dispersed in a comparatively gentle manner. On Venus, where plate tectonics either did not exist in the first place or seized up very early, the thick and continuous crust cannot disperse the planet's internal heat. On the contrary, the crust acts more like a stuck valve on a pressure cooker, building internal heat until the planet blows its top in periods of extensive volcanic activity. These epochs must flood the surface with oceans of molten lava and leave their signature in the thick and choking atmosphere of carbon dioxide that gives rise to the planet's infamous and horrendous runaway greenhouse.

On Earth, plate tectonics not only helps relieve the accumulation of internal heat but also makes the carbon cycle possible. Photosynthesis by land plants and phytoplankton near the surfaces of oceans takes carbon dioxide from the air and uses it in the synthesis of organic matter. Zooplankton consumes much of this material close to the ocean surfaces and, when their time comes to go the way of all flesh, sink to the ocean floor where their skeletons accumulate in a layer of carbonate and silicate material. Eventually, this material gets drawn down into the depths of the Earth by the subduction of the continental shelves. Meanwhile, on land rainwater (made slightly acidic by the presence of dissolved carbon dioxide) reacts with materials of exposed surface rocks, dissolving silica dioxide and carbonates and carrying these ions down to the oceans in rivers and streams. There they are incorporated into the skeletons of zooplankton and the shells of shellfish and corals, in the long run ending up on the ocean floor and (in the even longer run) carried into the depths of the planet by the continuous conveyor belt of mobile and subducting crustal plates. Deep inside the Earth, they are pressure cooked and the carbon dioxide released, returning to the atmosphere through volcanic vents and hot springs. This process has rightly been called the planet's thermostat, regulating global temperatures by regulating the amount of atmospheric carbon dioxide. This interplay of geologic forces and biology at the level of relatively simple organisms is something denied Venus by virtue of the nature of its crust, irrespective of its distance from the Sun.

The Earth is also fortunate in having a remarkably stable tilt to its axis of rotation. Sure, there are variations, about which more will be said in the following chapter, but these are slight when compared with those of Mars, for instance. If Earth wobbled over time to the same extent as Mars, dramatic climatic cycles would take place which, whilst probably not making androde life impossible, would at the very least make the progress of civilization more difficult.

The slight tilt in Earth's axis also aids in the shifting of major weather systems back and forth across tropical latitudes in an annual cycle of monsoonal rains. If our planet's spin axis was perpendicular to the plane of its orbit, this would not occur to such a wide extent and much of the tropical latitudes would be far drier. This, in turn, would mean a significant reduction in life's diversity at those latitudes and a general increase in the desert area in hot regions of the planet.

The speed at which our planet rotates is yet another Goldilocks factor. Rapid rotation means high winds, which would surely be detrimental alike to tall vegetation (trees) and tall animals (humans). We may also think that bipeds would be less suited to such a constantly high wind environment; walking on all fours lessens the risk of getting blown over! Atmospheric modeling indicates that rapid rotation also results in larger temperature differences between equator and poles. It is even possible that during the "snowball Earth" period which gripped the planet in pre-Cambrian times, polar temperatures may have become low enough on a very rapidly spinning Earth for carbon dioxide to freeze out of the atmosphere, resulting in a smaller greenhouse effect and persistently lower temperatures at later epochs. On a worse-case scenario, the Earth may still be frozen, with algae as the most advanced form of life.

On the other hand, very slow rotation implies long days and long nights, which in turn makes for very hot afternoons and very cold early mornings. It also means less temperature difference between equator and poles and an entirely different pattern of atmospheric circulation. If polar regions were as warm as the tropics, there would be no polar ice caps, the ocean levels would be correspondingly higher and the area of dry land reduced. Moreover, higher global temperatures and a larger air/water interface would lead to more evaporation, higher humidity, increased rainfall and, in consequence, higher rates of erosion for the "dry" land that did exist. We must also remember that, although carbon dioxide has acquired the reputation as the greenhouse gas per excellence, water vapor (and methane too, for that matter, although this is not relevant for the present) has even greater greenhouse potential. Could a significantly more watery Earth end up falling victim to a runaway greenhouse effect? Maybe, although the presence of extra cloud cover would need to be factored in as well. In any case, such a world looks a good deal less inviting than the real Earth.

In addition to these properties, Earth is also a good size for habitation by androde life forms. If it were much larger, the stronger gravitational field would probably make the degree of muscular agility and fast responses associated with the quick wits of smart land animals unlikely. Also, bipedal mobility becomes more difficult on a larger world. As it is, the price we humans pay for walking around on two limbs, even on the Earth we know, is a constant strain on our backs. A tendency to bad backs comes with the territory it seems, but we can live with this (though not always comfortably!) in a gravitational field of 1g. In more powerful gravity though, things would be much more difficult and painful.

Moreover, a super sized Earth at the distance of the real Earth from the Sun would probably turn out more like Venus, courtesy of a denser atmosphere and stronger greenhouse effect. Ironically though, these "Venusian" conditions might co-exist with very Earth-like plate tectonics, as some models of the likely evolution of super-Earths indicate that they are more likely to sustain plate tectonics than smaller worlds like our own. Earth itself is probably an anomaly in its possession of plate tectonics, apparently (in part at least) a legacy, as we said earlier, of its unusually thin crust. This is to some degree born out by the uniqueness of terrestrial plate tectonics in the Solar System. Super-Earths, which might also display this phenomenon, do not exist in our home planetary system. The only known examples of this class of planet are in distant systems around other stars.

At the other end of the spectrum, a smaller planet with a weaker gravitational field allows too much atmospheric gas to leak away into space to sustain a healthy ecosystem incorporating "higher" animals for any significant length of time. Moreover, planets somewhat smaller than Earth that might exist in so-called habitable zones today were probably not able to sustain a suffi-ciently rich biosphere through their early years when their suns were still in the "faint" phase. Stars like our Sun gave out less light and heat in their youth and, although Earth managed (just!) to hold enough heat to get through that time, smaller planets may not have been so fortunate. It used to be thought that carbon diox-ide in the atmosphere of ancient Earth saved our world from freez-ing thanks to strong greenhouse warming. However (as we will take up more fully in the following chapter) strong doubts have now been raised about this hypothesis. More recent thinking sug-gests that the oceans played a warming role by absorbing more incident sunlight than bare land, thereby warming the lower atmo-sphere of those early times. Planets just a little less massive than Earth are more prone to lose water vapor and are therefore not as likely to be covered by deep and extensive oceans in their early life. More bare land and less deep water means that more incident sunlight is reflected back into space without warming the lower atmosphere of these worlds. During the faint Sun era, that may spell frozen disaster.

Mention has been made in a couple of places about the great benefit that Earth's unusually thin crust has made to its habitabil-ity. In one sense, asking why we live on a rare thin-skinned world is a question which answers itself; if our planet had instead been thick-skinned, we simply would not be here to ask the question! A better inquiry however, is to ask why Earth has such a thin crust in the first place. The reason is an interesting one. Its developing crust was largely stripped off in a catastrophic glancing collision with a wandering planet about the size of Mars. This happened as Earth was still growing, around 4.5 billion years ago. The impacting planet hit ours at a glancing angle and disintegrated, its iron core crashing onto the young Earth and eventually sinking into and mingling with our planet's own. Much molten rock that would otherwise have become Earth's crust was blasted into space, together with remnants of the rogue planet itself. This material eventually come together to form the Moon.

On Earth itself, the impact's effects were dramatic. Any original atmosphere was blasted away (incidentally, saving Earth from a dense Venusian style mantel of gas) and the surface was transformed into an ocean of magma. So much energy was released by the impact that for thousands of years following the event, Earth sported a new atmosphere of vaporized rock which gradually rained out in drops of molten liquid. This was followed by a thick steam atmosphere lasting for about 10–30 million years. All the while, the impact spun up Earth's speed of rotation enormously, only to be slowed again by the tidal effects of the newly formed – and very close – Moon.

These tidal effects must have been horrendous. We can only imagine the frightful interaction between the powerful lunar tug and that nightmare magma ocean. The tidal stresses and strains added further heat to Earth, at the same time pushing the Moon away from our planet and slowing Earth's rotational speed to something closer to the "Goldilocks" velocity we know today.

After millions of years, in-falling volatile rich asteroidal and cometary fragments, together with the planet's own volcanic outgassing replaced the earlier atmospheres with one thick enough to permit the existence of oceans yet not so thick as to create a stifling heat blanket that would evaporate these oceans as the Sun warmed up over millions of years.

So where is the rest of Earth's crust today? If you want to see some of it, just look skyward on a moonlight night. Mingled with the remains of the impacting object, part of Earth's old crust makes up that shiny orb in the heavens. Other pieces presumably drifted away into space.

Actually, the Moon plays a bigger role in our planet's habitability than simply acting as a repository for some of the planet's unwanted crust. Several studies show that its presence acts as a tremendous stabilizing force that minimizes the long-term wobble in Earth's axis, maintaining a more stable climate over long periods than would otherwise have been possible. Kasting demonstrated that the Earth would need to rotate twice as fast as it does to maintain its level of stability, were the Moon not present, but as we have already remarked, rapid rotation means stronger winds and a less than congenial environment for taller organisms. Moreover, the climatic effects of a 12-h day do not sound very inviting. We will have more to say about the Moon's effect on the Earth and its role in maintaining relative climatic stability in the following chapter.

Other effects of the giant impact itself may be our planet's very favorable axial tilt and a greater concentration of heavy metals (from the impacting planet's core) than would otherwise have been possible. This latter, plus the extra spin imparted by the impact may also be at least partially responsible for another biologically very beneficial property of Earth, not shared by any of the other small planets of the Solar System. This is the relatively strong magnetic field which acts as a shield to the stream of particles constantly boiling off the corona of the Sun. Without this field, energetic particles would penetrate to the surface and the levels of stratospheric ozone might be stripped away faster than it could form. Serious ozone depletion would result in an increase in high-energy ultraviolet radiation at the planet's surface. An increase in both energetic particles and damaging UV would not be good news for land based organisms.

Major collisions are probably not rare in young planetary systems, but for one to have the effects that the Moon-forming impact had, the angle of impact needed to be just right. Furthermore, the time of occurrence was also constrained within a fairly narrow range. The results would have been different had the impact occurred after Earth had fully formed. Exact duplications of this event are presumably very rare and considering the importance that it is thought to have had for the existence of an environment suitable for androde life, we can probably conclude on these grounds alone that the chances of such life are very remote indeed. Put another way, had the rogue planet approached at a slightly shallower angle and missed Earth altogether, our planet would most probably have turned into a second Venus. Had it come in at a steeper angle and smashed straight into Earth, the "habitable zone" of the Solar System would now either be the site of a band of asteroids or, (if the material from both Earth and impactor came together into a single orb) a super-Earth with an atmosphere most probably quite Venusian in density and temperature.

It looks suspiciously probable that there is really no such thing as a "habitable zone" for androde life unless any planet within that zone has experienced a freak set of circumstances that has turned it into a cosmic oddity. Our own dear Earth, it seems, is just such as place and we are the living proof of that fact!

There is little doubt that before much longer – maybe even before you read these words – planets having Earth like masses will be discovered orbiting other stars. Some of these may even be located in the so-called habitable zones of these stars. But if what we have been saying here has any truth at all, that will not automatically mean that these planets harbor life, especially androde life. Sensationalist claims will no doubt be made in the media, but the truth is likely to be far more sober.

Enough has already been said to give some idea of the many factors influencing the conditions that might be found on planets that superficially look like our own. Now let's look at one more which raises some strange and interesting prospects in its own right.

A Diamond in the Sky!

Just as the Sun and its planets formed from material enriched in heavy elements cooked by previous generations of stars, so younger systems have formed from material having a different degree of enrichment to that of our own system formed. Some astronomers have argued that younger systems are likely to be more carbon rich and oxygen poor than our own and that planets which might be seen as counterparts of Earth and Venus would have very different internal compositions. By the standards of our familiar Solar System residents, these may look pretty exotic. Heat and pressure at the centers of planets of this size compress carbon into its crystalline form – in more familiar terms, into diamond. Not "diamonds" – plural – but into one enormous single diamond. These planets are expected to have diamond cores!

One planet richer in carbon than oxygen has already been discovered, but it is a giant having a mass some 1.4 times greater than Jupiter. This world (catalogued as *WASP-12b*) orbits a star 1,200 light years from Earth so closely that its "year" lasts a mere 26 h. Beneath the scorching top of its atmosphere, carbon in the form of graphite is believed to lurk, whilst deeper down diamond and maybe even more exotic forms of the element are thought to exist. Although this world is a giant, its very existence lends further credence to the presence of smaller carbon worlds, including ones of Venus/Earth mass.

Of course, their carbonaceous composition will have profound effects on the evolution of these worlds. The plate tectonics that have played such a vital role in both the climatology and biology of Earth depend upon a fluid core. We cannot expect this process to be evident on a world with a heart of solid diamond. The absence of a molten core would, moreover, take away the planet's ability to encase itself in a protective magnetic field. Although it cannot be categorically denied that life exists on such planets, it certainly seems that the environment there is a lot less conducive to the rich ecosystems of the type known on Earth and upon which our existence as a species depends.

As an aside, we may ponder the existence of even more exotic planets in these carbon-rich systems. In our Solar System, the planet Mercury has provided astronomers with some unexpected surprises. Before the advent of robotic interplanetary space probes, most astronomers thought that the Sun's closest planet was very similar to the Moon and Mars, i.e. a rocky world without a significant iron core. The planet was not thought large enough to be differentiated into a metallic core and rocky mantel and crust in the manner of the larger Venus and Earth.

Imagine the surprise then, when close flybys of Mercury revealed an uncommonly dense world largely comprised of a grossly outsized metallic core and a relatively thin overlay of rock! It is difficult to conceive how such a weird planet could even exist; how it could have been formed in the first place.

One hypothesis says that Mercury was not formed as it is today. The planet that now graces our twilight skies as the fainter of the two morning and evening "stars" is little more than the core of a far larger orb – "Proto-Mercury" if you like – that was broken apart early in the planetary system's lifetime through the impact of a large planetary body. Earth, as we have just seen, was also struck by a similar body, and we gained the Moon plus a thin crust and a moderate rotational velocity, all of which has aided in the appearance and preservation of life here. Mercury was not so lucky. The higher velocity impact that it experienced smashed it to smithereens or, at least, smashed the rocky outer layers to smithereens leaving exposed the metallic core. Some material from Proto-Mercury and probably from the impactor as well settled back onto the core, but most of the planet was scattered to the wilds of empty space. Hence, the strange little planet with an oversized core that we have today.

Now, the sort of giant impacts that shattered Proto-Mercury, gave Earth its Moon and also (it is thought) knocked Uranus over onto its side are not prerogatives of the Sun's retinue of worlds. They must happen in essentially all types of solar systems. We may legitimately imagine a planet like Earth or Venus in a carbon-rich system being struck and shattered by a large impactor. Like the Proto-Mercury event, some of these impacts could end up leaving just the core of the original planet intact; but what would this core be like in these instances? As we said above, the core would be a diamond! Systems in which such events have taken place must contain small planets that are solid diamonds. Imagine a diamond over 1,000 miles in diameter! Literally, a diamond in the sky!

The writer recalls a television play about a reporter (played by actor Lloyd Bridges) sent to cover the story of an astronaut launch and who, while waiting for the launch, fantasized about a distant future in which a group of astronauts landed on an alien planet, only to find that it was made of solid diamond. The story was intended only as pure fiction of course, but it is a little creepy to think that such worlds probably do exist and that it may not be too long before accurate extra-solar planet searches find them. Whether their true nature will be recognized or not is, however, another matter.

Is There Life Everywhere (Even if Androdes Are Oddities!)?

The tentative conclusion of this chapter may be disappointing to some readers. *Star Trek* enthusiasts might be happier if *Klingons* really did exist, and readers for whom the *Principle of Mediocrity* is the last word in scientific method might be wondering if old Copernicus is turning in his grave!

If the arguments presented here are right (always the "If" of course), human beings might be real oddities. Perhaps it is *we* who are the weirdest phenomena in the universe!

Yet this need not mean that life itself is rare and there might well be some truly exotic life forms out there on other worlds. We might not be able to communicate with them, but in their own way they could be every bit as strange as us.

Controversial cosmologist Sir Fred Hoyle and his equally controversial colleague Chandra Wickramasinghe, argued that life was "nature's preferred way" of synthesizing organic compounds of the type found not only in solar system material but also in interstellar dust. Such complex molecules even play a role in star formation, so if Hoyle and Wickramasinghe are correct, life has a truly cosmic role.

These radical cosmologists presented the following scenario. Cometary bodies form in abundance around newly formed stars. At least in the larger objects of this class, liquid water and carbonaceous compounds mix together in a warm soup deep within the solid body and it is within these concentrated warm pools that simple forms of life have their genesis. These are subsequently ejected into space through cometary activity and through impacts of meteorites on their home comets, where they become (either in a dead or dormant state) cosmic dust and, as such, contribute to the material from which a new generation of stars – and comets – form. Not all of these tiny organism get that far however. Some are swept up by planets, either as micrometeorites or deposited on planetary surfaces within the matrix of ordinary meteorites. Where conditions are suitable, these organisms thrive and eventually colonize the planet.

Now, very few scientists go as far as this, but the persistent (albeit controversial) claims of what appear to be simple microfossils in carbonaceous meteorites has kept alive the notion that the parent bodies of these objects may have maintained a solution of organic material and water long enough for some simple forms of life to appear and that, perhaps, Earth was seeded by ancient meteorites containing something more viable than microfossils. If it could be shown that at least some of these meteorites come from cometary bodies (broadly defined for this purpose as objects which shed significant quantities of dust and other material into space), it is arguable that Hoyle and Wickramasinghe might at least be partially right. Perhaps some of the organic material in the depths of space really does have a biological genesis. This is not as strange as it may sound. Comets pretty clearly abound in great numbers around most of the stars in the universe. If even a small proportion of these have at some time acted as sites of simple biological activity and if even a small percentage of these shed their load into space through the activity typical of such bodies, it would appear inevitable that *some* biological material ends up floating around in the void!

The type of meteorite chiefly suspected of having cometary connections is the carbonaceous chondrite, especially the more volatile-rich and less metamorphosed Types 1 and 2. Type 1 carbonaceous chondrites consist of the CI class while Type 2 is made of several sub-types which differ from one another in their detailed mineral content. The majority of Type 2 meteorites are the so-called CM chondrites, of which the Murchison meteorite that fell in Australia in 1969 is probably the most famous, and certainly the best studied, example even though the prototype is really the Mighei meteorite that came down in Russia in 1889. The "M" in "CM" stands therefore for "Mighei", not "Murchison" (nor "Murray", to name another famous meteorite of the same class that fell in Kentucky in 1950). The possible microfossils mentioned above are principally found in these Type 1 and 2 meteorites, but they have also been isolated from the more highly metamorphosed Type 3 carbonaceous chondrites such as the Allende CV3 which fell in Mexico in 1969. If the parent bodies of any of these meteorite classifications really do display cometary activity, the presence of biological material and even dead or dormant organisms in space may not be as farfetched as it sounds.

Unfortunately, most people who witness the fall of a meteorite are too overwhelmed by the spectacle of the phenomenon to take detailed measurements of its trajectory, so even tolerably accurate orbits of these bodies are few and far between. Yet, after a careful examination of reports of the CI meteorite Orgueil (falling in France in 1864), M. Gounelle, P. Spurny and P. Bland were able to calculate a number of possible orbits suggesting that the body orbited the Sun in an ellipse having an aphelion distance that was probably beyond the orbit of Jupiter, possibly beyond that of Saturn and maybe even out past Neptune. Their conclusion was that it was probably a fragment from a comet of the short-period variety (having aphelion near Jupiter's orbit) although one of Halley-type period was not altogether excluded. No specific comet is known in a similar orbit, but that is not surprising as any putative parent likely faded out or got deflected into a different orbit long ago.

The orbit of this meteorite is certainly suggestive of cometary origin, but that of itself is not proof of a cometary connection. It may be an asteroidal fragment flung into a comet-like orbit through a close encounter with one of the planets. For reasons which we shall look at in a moment, this possibility might be worthy of further thought.

Matching up orbits is not the only way to identify the parent objects of meteorites. Indeed, it is not even the best way as the gravitational interplay of planets scatters meteorite-sized fragments far and wide and into orbits bearing little resemblance to their origins.

Most attempts at matching meteorites with their parents involve comparisons between the mineralogical composition of the meteorite and that of an astronomical body as determined by the latter's reflectance spectrum. In this way, prospective parents (or, at the very least, parent types) have been found for a number of meteorite classifications.

It is here that the proposed identification of Orgueil as a cometary fragment rests on shakier ground than orbital computations alone might suggest. Certainly, the nuclei of comets are dark carbonaceous bodies, but most are distinctly redder in color than CI meteorites and therefore do not make very good matches to these objects. At least, this seems to be the case for what we might call "regular" or "classical" comets arriving on orbits stretching beyond that of Jupiter.

On the other hand, there is a much closer match between carbonaceous chondrites and asteroids of the C type, or some variation thereof. It is interesting to note that some asteroids of this classification have shown cometary activity (the so-called Main Belt comets) although none of the MBCs known to date has been named as a prospective meteorite parent. As C types however, these comets appear to be compositionally more akin to carbonaceous meteorites than the "classical" variety. Moreover, carbonaceous interplanetary dust particles reaching Earth as micrometeorites are found to have compositions very similar to CI and CM meteorites; and most of these particles have an asteroidal rather than classical cometary origin.

It is also interesting to note that water present in CI meteorites has a deuterium/hydrogen ratio similar to that of Earth's oceans, roughly half that of the water found in the comas of the few comets for which this ratio has been derived. This presents another difficulty for a proposed link between comets and CI meteorites (although it must be admitted that the data for comets is very small and does not include objects of very short orbital period), but it raises the very intriguing possibility that icy asteroids and something like Main Belt comets may have provided the original source of our planet's water. If CI meteorites are fragments of these objects, they may be the source of our planet's water as well as its organics and (if Hoyle and company are correct), the seeds of its life as well!

It is possible that Orgueil originated as a fragment of the Themis family of asteroids and Main Belt comets or, alternatively, a parent body forming from the same material as these denizens of the outer asteroid belt. It may or may not be significant that the possible orbits calculated for Orgueil have extremely low inclinations (less than 1°!) in common with the Themis objects. These bodies have inclinations less than 3° and some of them, like the meteorite, less than 1. Most carbonaceous dust particles reaching Earth are thought to originate from this asteroid family and some of these particles have carried organic structures not too unlike the controversial ones in the meteorites. Whether these are fossilized microorganisms spewed out of Main Belt comets is an interesting – if controversial – thought!

Also of great interest is a group of rare Type 2 carbonaceous chondrites. Initially grouped with the CM2s of the Murchison, Murray and Mighei variety, these are now given their own classification as CR2s. Only three have been observed to fall; the prototype Renazzo (hence "R") which fell in Italy on January 15, 1824, Al Rais (Saudi Arabia, December 10, 1957) and Kaidan (Yemen, December 3, 1980). No other fall of this type is known, although some old fragments have been recovered from Antarctica after having been preserved on ice for many years and a few have been found in dry regions where they fell long ago.

Some studies (albeit not undisputed as we shall see in due course) have paired these meteorites with asteroids of the rather rare B type; one of the sub-species of C type bodies. Initial studies indicated a very close match to 2 Pallas and, even more so, to the small asteroids of the Pallas Family which are almost certainly fragments scattered from Pallas itself following the violent impact of another body in the distant past. Recently, researcher H. Campins of the University of Central Florida demonstrated how small asteroids broken away from Pallas may also be deflected into Apollo-type orbits venturing well within the orbit of Earth. One such object which, according to Campins, probably originated in this way is the strange little asteroid 3200 Phaethon. Like Pallas and its Main Belt progeny, Phaethon is also a B type, but it has always been something of a headache for astronomers by being the parent body of the most intense and spectacular of all the annual meteor showers – the Geminids. Meteor showers are normally associated with comets, and are easily explained as arising from relatively large particles ejected from the cometary nucleus through the sublimation of ices. Yet, Phaethon had not given any indication of cometary activity and its very small perihelion distance of just 0.14 AU made the long survival of ice beneath its surface problematical. Nevertheless, dynamical studies of the Geminid meteors reveal them to have been ejected near perihelion and not, as an alternate view had suggested, from a collision between Phaethon and another small asteroid close to the former's aphelion in the asteroid belt.

(3200) Phaethon (Geminid parent body)
25 December 2010, stack of 4 x 150 seconds (20 minutes spaced)
Marco Langbroek, using remote 37 cm F14 Cassegrain, Winer Observatory Sonoita (MPC 857)

Fig. 1.5. The "Geminid asteroid" 3200 Phaethon tracks through this star field on Christmas night 2010 as imaged by Marco Langbroek using the 37 cm. F14 Cassegrain telescope at Winer Observatory, Sonoita. The image is a stack of four images of 150 s exposure each (Credit: Marcoaliaslama)

Then, for a few hours when near its perihelion on June 20, 2009, Phaethon unexpectedly brightened by at least two magnitudes in a brief outburst of weak comet-like activity. A study by D. Jewitt and J. Li interpreted the brightening as due to a release of dust and suggested that, because of the strong surface heating when the asteroid is near the Sun, dust is produced by the cracking of hydrated surface materials, somewhat analogous to the process known as "exfoliation" of terrestrial rocks in desert regions where expansion and contraction between the extremes of daytime and night time temperatures causes the surface of rocks to flake off in leaf-life (hence ex*foli*ate) sheets. On Phaethon, according to Jewitt and Li, dust produced by this decomposition process becomes elevated above the surface of the asteroid by electronic suspension and subsequently swept away by the pressure of sunlight. We will, incidentally, see more of the electrostatic suspension of particles in Chap. 6, where it is examined as the cause of an unexpected class of phenomenon witness by astronauts on the Moon – but more of this in due course.

Jewitt and Li call Phaethon a "rock comet"; a body capable of a degree of activity even in the absence of icy material. It is quite likely that Pallas contains ice underneath its surface and may therefore have experienced bouts of cometary activity, especially at the time of separation of the other members of its family. The relatively low density of the asteroid suggests that quite a lot of ice might remain within it even to this day. Following the long ago disruption that gave rise to its family, Pallas and its progeny (including Phaethon) probably exhibited cometary activity for a time. Phaethon may have retained enough ice to have activated again as its orbit evolved into the small perihelion ellipse of more recent times, so maybe it was more than simply a rock comet initially. There is, in fact, some evidence supporting this view. Two tiny asteroids, each accompanied by its own weak meteoroid stream, are known in orbits not unlike that of Phaethon and it is thought that these are fragments that split away from Phaethon thousands of years ago, likely as the result of an explosive release of vapor from sublimating internal ices. At that time, Phaethon presumably suffered a large cometary outburst, and it is also likely that both it and its two progeny continued a level of activity through several further perihelion passages until the freshly exposed ice evaporated away.

The nature of the Geminid shower also indicates that something more than the very low level of activity displayed in 2009 was once the order of the day. Some of the Geminid fireballs are caused by objects the size of house bricks and it does not seem possible for electrostatic suspension and solar radiation pressure to lift objects of that dimension and expel them into space! It is therefore tempting to see this as evidence of former cometary activity, but the difficulty in that neat suggestion is the age of the Geminid stream. This is thought to be only around 1,000 years (give or take a few centuries) and as such formed rather late in Phaethon's career, presumably after any ice had sublimated away into space and sublimation-driven cometary activity long ceased. It is tempting to suggest that liquid water, no doubt trapped in rock pores or in the form of thick mud, might exist within Phaethon even today and occasionally erupt from the surface through fresh cracks.

Perhaps the majority of Geminid meteoroids were blown out in a kind of geyser!!

Yet, regardless of what activity Phaethon may or may not have displayed in the past, it does appear that if this object and CR2 meteorites both come from the Pallas stable, these meteorites are also linked, at least indirectly, with an object that continues to shed dust into space. This implies that if this object is indeed composed of Type 2 carbonaceous chondritic material and that if Hoyle and colleagues are correct in their assessment of some of the organic compounds and structure therein being of biological origin, then maybe some of the stuff being puffed off the surface of Phaethon really is biogenic! Even if Hoyle and colleagues went too far in their speculations, it now seems that the organic contents of CR2 meteorites are of even greater biological interest than those found in objects like Murchison. That at least is the conclusion of a 2010 paper by Sandra Pizzarello and Everett Shock. The following year, Pizzarello and her University of Arizona team announced that they had also extracted ammonia from a CR2 fragment found on the ice in Antarctica. Even if life itself was never present in these bodies, much of life's chemistry certainly was.

It is possible that some meteorites are directly related to Phaethon. Campins points out that because of the relatively slow velocities of the Geminid meteors (and, we might add, their relatively high density when compared with meteoroids of many other showers, plus the considerable size of the shower's largest members) the Geminid shower is one of the few which may be capable of delivering meteorites to Earth's surface. The central core of the Geminid shower extends from December 6 to December 19, with the peak period being around December 13–14. However, traces of the shower have been detected as early as November 30 and as late as December 29, so that some level of Geminid activity is present throughout essentially the whole of December.

The link between Phaethon and the Geminids is about as firmly established as anything of this nature can be. The Phaethon/Pallas link is not as strong, but still looks good. The Pallas/CR2 link may not look as good as it once did however. A study published in 2010 appeared to find a closer resemblance between B-type asteroids and more highly altered carbonaceous chondrites (including Allende-type CV3s) than with CR2s, although the

Fig. 1.6. A brilliant Geminid fireball lights up the Mojave Desert at Hercules Finger rock, near Victorville CA, as captured in this photograph by Wally Pacholka on the morning of 14 December 2009. © Wally Pacholka/AstroPics.com/TWAN

density of Pallas seems closer to the latter. If the CR2 connection continues to hold, the Geminid meteoroids should be of that composition. This is consistent with the results of studies of Geminid fireballs indicating a density similar to that of CM and CR meteorites, rather than with CV types.

Now, is it just coincidence that two of the three known CR2 meteorite falls occurred during the period when the Geminid shower was active?

The Al Rais meteorite fell as the shower was working up to maximum. Unfortunately, the writer has been unable to find any description of its trajectory or even the time of day when it fell, so any possible association with the Geminids rests solely upon the calendar date and the proposed CR2/Pallas/Phaethon association.

A little more is known about the fall of the Kaiden meteorite. It fell around 7.45 A.M. local time on the morning of December 3 and the fireball was said to have been traveling from the northwest to the southeast. The apparent speed of the fireball was not given, nor was its elevation at the time it first became visible, but it is noteworthy that the constellation of Gemini was in the north-western sky at the time of the fall, so what we do know about the meteorite's arrival appears consistent with the fireball having been a Geminid.

It should be mentioned that some researchers have proposed the Martian moon Phobos as the parent for this particular meteorite. That in itself would be interesting, as this small moon is believed to contain ice and some of the pits in its surface, spied by Mars-orbiting spacecraft, may have been caused by eruptions of gas in times past. Russian spacecraft have, moreover, found evidence of gaseous emissions from this moon, so we may well wonder how comet-like *it* would be if it passed closer to the Sun!

Be that as it may, the Pallas-via-Phaethon path here proposed for the Kaidan meteorite is at least superficially strengthened by the trajectory of the fireball away from the general region of Gemini. The evidence for the Kaiden and Al Rais meteorites being Geminids might be circumstantial, but people have been hanged on less! You be the jury.

Project 1: Observing the Geminid Meteor Shower

As well as being of great interest because of its association with Phaethon and (indirectly) Pallas, the Geminids are also one of the most rewarding annual meteor showers to observe.

The shower was discovered independently by several observers in 1862 and reports throughout the remainder of the nineteenth century indicate that it provided only a comparatively weak display in those days. The level of activity of meteor showers is measured by the Zenithal Hourly Rate (ZHR) or the number of shower meteors seen by an observer with a clear, dark sky in which stars of magnitude six are visible overhead and with the radiant (i.e. the small area of sky from which the meteors appear to radiate) in the zenith. Early reports indicated a Geminid ZHR of around 14. By the late

(continued)

Project 1: (continued)

1800s, this had climbed to the low 20s and remained close to this value until the 1930s, when the rates jumped to between 40 and 70. These leveled out to around 60 during the 40s and 50s, climbed a little during the 1960s when the shower rivaled the more famous August Perseids. During the 1970s, ZHRs climbed into the 80s and rates between 1980 and 1985 ranged from around 60 to as high as 110. By the early 2000s, rates for this shower varied from around 120 to 160, making it the strongest of the annual meteor displays.

This is not because more particles are being fed into the stream. It is just that the orbit of the stream slowly changes over time and Earth now encounters a denser region than it did in the 1800s. The shower will likely continue to increase for several years into the future, but eventually it will dwindle off again and in centuries to come, cease to encounter Earth altogether.

The Geminids can be observed from nearly every part of the Earth and the radiant is well marked by the two bright stars Castor and Pollux. To observe the shower, wait until these stars are high in the heavens, preferably about 2–4 A.M. The paths of the meteors, when extended backward through the constellations, appear to converge close to Castor. This is the radiant of the shower, but that does not mean that the meteors are confined to this region. They can appear anywhere in the sky and the further from the radiant, the longer will be their apparent paths.

The shower has quite a sharp peak of activity around December 13–14, with hourly rates remaining around half the maximum for about 2 days. During the maxim, an average of two or three meteors per minute can be expected for a single observer blessed with good skies and the radiant high overhead. But the meteors will not come in a steady stream. Meteor showers tend to produce short bursts when the rate (if translated into ZHR values) soars well above average, interspersed with below average quiet periods. There will probably

(continued)

Project 1: (continued)

be times when you see several meteors simultaneously or in very quick succession, followed by several minutes without activity. Although most of the meteors will not be especially bright, always be on the watch for the odd fireball.

Most Geminid meteors do not leave luminous trains. From your observations, you might like to determine the percentage of the observed meteors that *do* leave trains. Are these mostly the brighter ones? Is there a magnitude cut-off where trains no longer occur? Conversely, is there an upper limit of brightness where trains always happen? (You will probably need to observe the shower over several years before you can even come close to answering this one!)

These are only a few things to watch out for, but most importantly, just enjoy the show!

Before leaving the subject of Pallas and asteroids that double as weak or part-time comets, some very interesting observations made long ago by none other than William Herschel and J. Schroeter should be mentioned. Both of these astronomers reported observing nebulosities surrounding the asteroids Pallas and Ceres. According to their descriptions, the nebulosities were extended and subject to frequent changes, which would seem to make them more like cometary comas than planetary atmospheres. Herschel also observed a very small nebulosity surrounding asteroid Juno, another of what are sometimes referred to as the "Big Four" asteroids, that is, the first four to have been discovered. Nothing was seen surrounding Vesta, the remaining and visually brightest member of the original quartet.

Although these observations were accepted by some astronomers during the nineteenth century, by the twentieth they were mostly rejected as mistakes. Similarly, a handful of observations by other astronomers of apparent comas surrounding asteroids were likewise rejected. Thus, J. Comas Sola's reports of nebulosities surrounding 224 Oceana and 182 Elsa were dismissed after other astronomers failed to confirm them and an astronomer who found an apparent 3–5 arcsecond nebulosity surrounding a photo-graphic image of 899 Jocasta later admitted that he found the observation insufficiently interesting to report! Yet, it is not easy to believe that astronomers as experienced as these could have been mistaken.

More recent years have witnessed several developments which cast a new light on these discredited old observations.

Firstly, on December 18, 1977, M. Lovas discovered an appar-ent comet on a photographic plate exposed at Konkoly Observa-tory in Hungary. He relocated the object on January 2 and 5 and on each occasion described the object as diffuse with central conden-sation. A further observation at Harvard on January 16, however, revealed only an asteroidal point of light. Calculations by B. G. Marsden revealed the object to be moving in an orbit of the Pallas type and later analysis indicated that it was also a B-Type similar to 2 Pallas itself. It is now known as asteroid 3579 Rockholt, but what are we to say about Lovas' indication of an apparent coma? Apparently, he later concurred that the object appeared "essen-tially stellar" in his January images, but seems to have held to his description of the December discovery images as being diffuse. As an experienced observer and discoverer of both asteroids and com-ets his opinion is not to be lightly glossed over. Interesting also that this object appears to have begun life as a fragment of 2 Pallas and as such is presumably a sibling of Phaethon.

Secondly, in 1990 Michael A'Hearn found evidence at ultra-violet wavelengths for a thin envelope of OH ions surrounding Ceres. Originating from the photodissociation of water molecules, the presence of this "UV coma" implies water in some form on the asteroid. Hydrated minerals constitute a possible source, but frost has been suspected on the surface of Ceres and it is widely thought that quantities of ice and maybe even liquid water might exist deep within this body. Perhaps the asteroid is given to the occasional eruption of water vapor from its deeper regions, causing it to become surrounded briefly by a cloud of fluorescing gas sufficiently bright for the likes of Herschel & Schroeter to observe from Earth.

Thirdly, we may note the steady stream of Main Belt comets turning up in discovery statistics since the mid-1990s. During their inactive phase, they look just like any other asteroid. We may well wonder about the converse of this: How many other "ordinary asteroids" have the same capabilities but have not displayed them – or have simple been *missed* displaying them – in recent decades?

Fourthly, asteroid 596 Scheila dramatically burst forth into activity in December 2010, growing an obvious coma and forked tail that each persisted for several weeks. The spectrum of the coma and tail revealed only reflected sunlight, so the outburst was more likely the result of an impact than of ice-driven cometary activity, but the sudden and unexpected nature of the event and the very comet-like appearance of the dust cloud at least had the effect of reviving some interest in the old reports.

And then there is Phaethon with its Geminids and weak little burst of action in 2009!

Far from being the celestial vermin that some astronomers have reckoned them, asteroids may turn out to be interesting and unexpectedly important denizens of our strange skies and may have played an unexpectedly large role in making our planet the blue oasis that we know today.

Project 2: Asteroidal Comas?

Some appear to be the products of meteorite impacts, some are definitely triggered by sublimating ices and some may result from "rock cometary" activity. Some may combine one or more of the above (an impact exposing sub-surface ice for instance). Whatever the cause, asteroids do at times puff out comas and even grow tails, and a careful watch might catch one in the act.

Amateur astronomers with telescopes of moderate aperture can keep watch on the brighter asteroids, but a serious program, of asteroid monitoring would be a better project for a number of advanced amateurs equipped with CCD technology capable of monitoring large numbers of the fainter bodies. Although a collision-driven coma can erupt from any asteroid at any time, asteroids having spectral types closer to cometary nuclei (dark objects related to the C-types, e.g. D, B, G, T, F and, of course, the various classes of Cs themselves) may be the best ones to monitor. As these comprise most of the members of the outer Main Belt, there are plenty of choices!

(continued)

Project 2: (continued)

A serious monitoring project is probably best conducted by a network of amateur astronomers, each member having been assigned a list of asteroids to observe at every opportunity. A large enough network capable of monitoring numerous asteroids might uncover something very interesting!

Things to look for include any sudden and unexpected brightening or a slightly nebulous or non-stellar appearance of the asteroid (Scheila appeared slightly non-stellar before its coma became properly discernible) and, of course, any sign of a surrounding nebulosity or tail. The latter may or may not be orientated in an anti-solar direction and may not necessarily be accompanied by an obvious coma surrounding the asteroid itself.

If any activity is suspected, other members of the network should be asked to confirm it and, if verified, a report made to the Central Bureau for Astronomical Telegrams at Cambridge, MA.

Appendix: Life in an Infinite Universe (A Brief Excursion into Philosophy, for Those Who Like to Ponder Such Issues!)

(The following has been relegated to an Appendix of this chapter because it raises issues which, though obviously related to the Chapter's topic, take us into areas which not everyone may wish to enter on a first reading of this book. If you find the subject too farfetched, go on to the next Chapter and – if you wish – return to this Appendix at a letter time.)

The question "Is the universe finite or infinite?" is one that has long vexed cosmologists, as well as one that raises all manner of philosophical issues, some of which bear directly on the question of androde life. It is frequently argued that, if the universe truly is infinite, it is certain not merely that other androde beings exist, that that there must exist infinite exact copies of the Earth and its inhabitants. The reason for this astonishing conclusion will become clear shortly.

The issue of a finite vs. an infinite universe appears to be a choice between two alternatives that both come with their fair share of difficulties, conceptual as well as physical. How can we imagine an infinite universe? Yet, how can we imagine a finite one? If the universe has a boundary of some sort (which is what "being finite" implies) what lies beyond it? And if something does lie beyond it, how is that "something" not another part of the universe?

The latter issue can be answered in a sort of way by appealing to some familiar "boundaries" which do not imply a "beyond." The popular one is the North Pole. This is the "boundary" of absolute "north-ness" for Earth's surface. There is nothing north of the North Pole, not in the sense that there is an impenetrable barrier there with a great emptiness beyond, but in the sense that there is no "beyond" at all! Similarly, to say that there can be nothing smaller than a geometric point does not imply that any dimension smaller than a point is empty of content. On the contrary, it means that there is no smaller dimension in which any content can be placed.

Infinity brings a different set of problems. We cannot imagine something that goes on forever, but that in itself is no objection. Nature is not, after all, limited by our imaginative abilities!

Traditionally, a universe which has a flat or Euclidean geometry is considered infinite. The same applies to one which has an hyperbolic or Lobacheveskian geometry. On the contrary, if the geometry of the universe turns out to be closed or Reimannein, it will be finite, although unbounded, rather like the surface of a sphere (except that this sphere has a three dimensional "surface" and exists in four dimensions!). We need not worry about these three classic possible geometries, except to say that all evidence points to the real universe being very nearly flat or Euclidean, without absolutely ruling out either of the remaining alternatives at the largest scales.

More recently, a few mathematicians have raised questions of shape or topology and suggested that even a Euclidean universe may be finite if its overall shape is analogous to that of a football, i.e. a ball consisting of a series of "faces" each having a flat geometry. A second group of mathematicians proposed an alternative topology; the so-called Picard topology in which the universe is shaped like a filter funnel. The universe that we observe is located, according to this model, around the flared end of the funnel. If we could travel deeper into the funnel, we would find that space became increasingly curved until it totally folded up into a one-dimensional line of infinite length (so infinity is not *quite* eliminated in this model). Near the pint where spatial curvature became one dimensional, we could imagine a region of space where the curvature changed so quickly that, were it possible for children and writing desks to exist under such circumstances, two school children sitting at adjacent desks and drawing triangles in a geometry class would get two different answers for the sum of the angles in a triangle; and both would be correct! Neither answer, by the way, would be 180°, as this only applies to triangles in the effectively Euclidean space way out near the mouth of the funnel!

Whether either of these topologies turns out to be correct for the real universe remains to be seen. Observational evidence for the football-shaped topology is possible if the diameter of the "face" in which we live is smaller than the distance to the most remote measured galaxies. This sounds bizarre, but it is possible. It means that when we look out into the universe, we are seeing the same set of galaxies over and over again. Nearby galaxies – say, up to a few tens of millions of light years away – are observed as they were when the light left them, in the relatively recent cosmic past. But (on this model) what seem to us to be more distant galaxies are simply the closer ones seen as they were at ever more remote epochs of the past. The universe begins to look like a hall of mirrors! Groups of astronomers have attempted to test this model by seeking repeating patters among groups of galaxies, but nothing conclusive has been found to date. Of course, the basic topology may still be correct, but if the diameter of the plane exceeds the distance of the furthest galaxies, observational evidence will be very difficult to find.

Topology may yet avoid the infinity issue, but if it does not, and if the universe really turns out to be Euclidean and infinite in extent, some weird consequences appear to follow.

Even if the universe at large is infinite, the universe as we observe it is clearly finite. As we look out into space and back into time, we see a younger and younger cosmos until, eventually we arrive at the wall of cosmic background radiation marking the release of the first light from the dense primordial sate of the universe. This took place about 300,000 years after the Big Bang. Because the universe is finite in age, even if infinite in extent, what we observe as *our* universe has boundaries. Indeed, even if the universe had been of infinite age, as long as space was expanding, we could not see out to infinity. We would still reach a boundary where the apparent recession of galaxies equaled the speed of light. Beyond this, nothing could be seen.

This observable universe – "our" universe – is sometimes called the O-sphere or observational sphere. An infinite universe may be looked upon as an infinite set of overlapping O-spheres. Now, this is where things start getting weird. According to quantum physics, space is not really a continuum, but is "grainy" at levels of the so-called "Planck length." This is not the length of a piece of wood, but the smallest dimension having physical meaning, named after Max Planck, the father of quantum theory. It is a mere 1.6×10^{-35} m, too minute for any direct experimental verification. Nevertheless, the grainy nature of space means that there is a limit to the number of configurations that can exist. Even if the universe really is infinite in extent (even if the number of O-spheres is truly infinite) there will still be a finite (albeit very large) number of differences that can exist between them. The weird conclusion is that *each O-sphere will be infinitely reproduced*! There will be an infinite number of O-spheres exactly like the one in which we live. This means that there will be an infinite number of Earths exactly like ours, even peopled by an infinite number of people exactly like us, thinking exactly the same thoughts and doing exactly the same things as we do. But if that is not weird enough, there will also be an infinite number of not-quite exact copies of our O-sphere. There will be infinite O-spheres exactly the same as ours except that Mr. Smith wears a hat today whereas in ours he wears a cap. There will be an infinity of O-spheres exactly like this one, except that the writer's hair turned grey with age instead of falling out. There will be an infinite number exactly like ours except that the rock which today falls off a cliff on the fourth planet from the millionth star in the outer arm of the Andromeda Galaxy manages not to topple until tomorrow, and so on and on!

If this is true, a strange paradox emerges, which is best exemplified by a little science fiction tale. Deliberately overlooking the inconvenient absolute speed of light restriction, let's imagine a Captain Kirk of the remote future navigating a star-ship which cannot only zip at super warp speed around the galaxy, but even flit through entire O-spheres with the ease that a Saturday night joy rider speeds through neighboring suburbs. Captain Kirk leaves Earth orbit and heads out where no man has ever gone before. Picking up speed, he leaves our galaxy, then our local group of galaxies, then our O-sphere. He traverses O-sphere after O-sphere, looking for the nearest that is an exact clone of our own. After a certain time has elapsed, he finds it (don't ask how, this is only a fictional story!) and with a little more searching, hones down on the Milky Way look-alike and finally on Second Earth. He arrives to a heroes welcome! His friends and family rush over to him, welcoming him back home.

"But" he stutters, "I have not been here before. I left another Earth in a distant O-sphere and have traveled- across the vast cosmos to this exact duplicate of my home planet. The man you think you are welcoming is not me, but an exact counterpart of me."

"Well" someone says, "let's test this. Do you remember me from college?"

"Well, you do look a lot like Joe Black."

"That Jim, is because I am Joe Black! Remember the time we played a trick on the physics tutor?"

"Oh yes, we put an anti-gravity device in his pocket. Took him an hour to come down off the ceiling."

"That's right Jim. It happened just like that."

And so on. Every memory Captain Kirk has of his life is confirmed by his friends on this "second" Earth. You see, as "our" Captain Kirk left "our" Earth, "their" Captain Kirk left "second" Earth and would now be experiencing exactly the same conversations after landing on "our" Earth as "our" Captain Kirk is having on "second" Earth!

Confused?

Well, try this thought. In what way can "our" Captain Kirk and "their" Captain Kirk be differentiated?

Two identical twins can be identified by reference in one way or another to the framework in which we experience them. One is inside the house and the other outside, to use one example. But the ultimate framework for the whole observable universe is the O-sphere, and what makes the "two identical Captains Kirk" situation so awkward is that not only are the Captains duplicated, but the O-sphere – the ultimate framework itself – is duplicated! Not only duplicated, but replicated infinitely; with one feature of each identical O-sphere being Captain Kirk leaving his O-sphere in search of its identical counterpart!

The fact of the matter is, there is no way of observationally differentiating between the identical O-spheres. Does it therefore make sense to speak of *two* – or an infinite number of – identical O-spheres at all? As far as Captain Kirk is concerned, his experience could be just as easily interpreted as his having circumnavigated a spherical universe and come back to his starting point. These two alternative interpretations equally explain his observations; in other words, they are indistinguishable. So are they each, in a sense correct?

The philosophical issue here is known as the "identity of indiscernibles" or, in plain English, if A and B have exactly the same properties (including relational ones) does it make sense to speak of them as different entities?

The immediate reaction to Captain Kirk's identical O-spheres is to say "Of course they are different. The two are simply exact copies of each other." But sense can only be given to "A is a different object from B" if there is some way of differentiating between them. In the case of Captain Kirk's O-spheres, if every object, every event and even every thought is indistinguishable, all means of differentiation are *ipso facto* ruled out!

This may be even more apparent if the universe is looked upon from the point of view of information content. From this point of view, the basic unit of the universe is not a subatomic particle but a bit of information. The O-sphere is, essentially, a great store of information. Therefore, it could be said that the O-sphere which Captain Kirk leaves and the one in which he arrives are identical because the store of information is identical in each. Our intrepid space captain is simply re-encountering the exact store of information that he previously left behind! Considered in that light, the universe is (in a sense) both infinite and finite at the same time. (That need not be as paradoxical as it sounds. Draw a 3-in. line on a piece of paper. The line is finite. Right? Of course it is, but do you remember having been taught at school that a line consists of an infinite number of points? So congratulations! You just drew an infinite line – infinite, at least, from the perspective of a single point!!!)

It may be worth mentioning here (though, mercifully, it will be only a brief mention!) that some interpretations of quantum theory rely on the existence of "many worlds" or alternate universes, not scattered through an infinite space, but as alternate spaces existing "alongside" our own. Interpretations of quantum physics are a little like alternate universes themselves, and we stress that the so-called "many worlds" interpretation is just one of the list, but the only point in bringing it up here is to point out that most proponents appear to interpret it not so much as an infinite assemblage of "other" universes as an assemblage of other versions of this universe. It is a rather strange psychological fact that where alternate universes are proposed in quantum theory, they tend to be presented in this light, while alternate O-spheres are nearly always expressed as truly "other" more or less similar clones of the known universe.

But is all this talk of infinity and "other" universes (however we might get our minds around these things) so much nonsense?

Infinity is at best a slippery subject, and many scientists who dare to cross the threshold into philosophy are not entirely happy with it. We could say that any concept that raises the sort of discussion as the above has got to have something wrong with it, and that is in effect what many philosophically inclined scientists are saying. Mathematician Jana Levin, one of the topologists who proposed the football-shaped universe, raised the point that if a mathematician finds that an equation gives an answer of infinity, he immediately concludes that something is amiss with the math. Why then, she asks, when the "equation" of the universe gives an answer of infinity according to certain models, is this answer accepted at face value and not as a warning that something, somewhere, has gone wrong with the theory?

Then there are the paradoxes raised by the amazing "Hotel" invented by mathematician David Hilbert. Imagine a hotel with an infinite number of rooms. Last night, an infinite number of guests arrived, and were immediately found accommodation. Today, an infinite number of other guests arrived and, despite all of last night's guests still being in the hotel, an infinite number of empty rooms still await the news guests. Despite the infinite guest intake last night, there were still as many rooms awaiting today's guests. Moreover, even after both lots of guests settled into their rooms, the hotel still has an infinite number of vacancies!

If that is not paradoxical enough, suppose now that an infinite number of the guests decide to go elsewhere. How many rooms are now left vacant? The answer, an infinite number; exactly the number that were vacant before the infinite number of guests left. And how many guests is the hotel left with after an infinite number leave? An infinite number of course! The numbers haven't changed!

This type of reasoning led philosopher John Anderson to flatly deny that infinity is a *number* at all, despite mathematical arguments to the contrary. Anderson saw it, in effect, as a way of saying that the universe never ends; that one can never come to the end of it, simply because there is no "end of it"! It is an *asymptote* rather than a true number with a set value. Still, I don't think that Hilbert's Hotel Paradox is solved simply by denying that infinity is a number. Read through the above several paragraphs again but replace "an infinity of ..." for each occurrence of "an infinite number of ..." (as Anderson would suggest that we do) and see if you think the Paradox is made less paradoxical. It does appear a little more subtle but the problem remains, I think.

The real issue is whether it makes sense to speak about "infinity" at all, whether as a number or as anything else. The universe is certainly large, and appears infinite for all practical purposes. But "infinite for all practical purposes" is not the same as being infinite in the absolute sense, in which case the bizarre paradoxes at which we have been briefly looking remain intellectual exercises only.

2. Our Changing Planet

Raindrops in Hell?

Not long ago, the early Earth was imagined to be a world of molten-rock oceans and belching volcanoes. Those of us who were around in the 1950s can probably recall the images in Walt Disney's *Fantasia* of a planetary ocean of lava bubbling away to classical orchestral rhythms. In the words of an early 1960s work by popular science writer Poul Anderson, "… we can be sure that for millions upon millions of years Earth was a fearsome, shuddering, flame-spurting Tartarus." Another writer several years earlier even went so far as to say that, if viewed from outer space, the Earth in its infancy would have resembled a miniature version of the Sun!

This very early period of our planet's infancy was called the "Hadean Era." "Hadean" means "hellish" although strictly speaking the term "Hades" only means the world of departed spirits, not necessarily "Hell" in the narrower – and hotter! – sense. Maybe, following Anderson, the period should have been called the "Tartaric" era, or something similar. In any case, this earliest period was supposed to have been hellish indeed. True, the ocean of molten rock and the flame-spurting image did not remain unmodified, but only by being replaced by a world turned into a hell by the constant influx of colliding meteorites.

Very early Earth was undoubtedly a hell of a place, but a discovery in Western Australia last decade set people reconsidering the degree of hellishness of at least the later period of the Hadean Era. No rocks survive from that distant age, but some of the oldest yet found from the subsequent era contain crystals of zircon whose date places them squarely within Hadean times. These crystals – discovered within very old rocks found in Western Australia – were formed at least 4.2 billion years ago, at a time

D.A.J. Seargent, *Weird Weather: Tales of Astronomical and Atmospheric Anomalies*, Astronomers' Universe, DOI 10.1007/978-1-4614-3070-4_2,
© Springer Science+Business Media New York 2012

when most scientists thought Earth was still in the throes of violent upheavals. But here lies the surprise. Oxygen isotopes trapped within these crystals yield important information about the conditions under which they were formed. Because the ratios of these isotopes are sensitive to the temperatures existing at the time of their formation, as well as to the presence of water, determining these ratios can tell us much about the prevailing conditions of that remote epoch. The unexpected result was that they indicate a relatively cool and wet prevailing climate. Earth's early "Hell" was cool and wet during its later periods! This, we remind ourselves, refers to a time earlier than 200 million years after the Earth itself condensed from the solar nebula.

Cool Sun, Warm Earth!

According to the zircon results, Earth was not a bubbling ocean of lava back in Hadean times. But if the picture of an Earth made hellish by internal volcanic forces and colliding meteorites is a massive exaggeration in one direction, an opposite picture of a world of eternal ice must also be explained away. Ironically, that is the picture painted by astrophysical theory as what "should" have prevailed then. Why? Because the theory of stellar evolution predicts that around four billion years ago – at the end of the Hadean era – the Sun was giving out just 70–75% as much energy as it is today. This is sometimes rather loosely termed the "faint Sun paradox." It should more accurately be called the "cool Sun problem" as a Sun that was 25–30% cooler than that of today should have meant a very much cooler planet than the Earth we know. Yet, from all available evidence that was not so. As well as the zircon evidence that helped put out the fires of the supposed early Hell there is ample evidence from the most ancient terrestrial rocks to prove that, far from being a frozen wasteland, the ancient Earth was awash with liquid water. For this to be possible, temperatures around about the same as those experienced today must have prevailed during the period immediately following the Hadean era, as well as during the later phases of that era itself.

How could this be possible? How could a cool Sun manage to shed as much warmth on our planet as the far brighter Sun of today?

The obvious answer is for the Sun's warmth to be conserved in some way. Imagine a camper with a blazing fire on a cold night. If he wants to remain warm throughout the night he must either keep the fire blazing high, or wrap himself in blankets as the flames die down. He must either keep the radiant heat constant or do something (wrap up in blankets) to stop the heat that he has received from escaping back into space. The early Earth had no choice in the matter; the Sun was cool and would not significantly warm up for millions of years. Somehow it had to conserve the relatively little heat being shed upon it. But how was this done? What acted as Earth's "blanket"?

A popular hypothesis relied upon the greenhouse effect, especially that caused by allegedly high levels of carbon dioxide in the early atmosphere. The so-called greenhouse gases (such as carbon dioxide, water vapor and methane) allow electromagnetic waves of relatively short wavelength to penetrate the atmosphere and be absorbed into the ground. This raises the temperature of the surface and lower air, and this energy escapes back toward space in the form of infrared ("heat") radiation, i.e. electromagnetic radiation having a longer wavelength. Greenhouse gases are simply those that are more or less opaque to waves of this longer frequency. They prevent heat radiation from escaping back into the cold sink of space. Instead, it is absorbed by the greenhouse gases, in the process raising the temperature of the lower atmosphere.

For this mechanism to work, early levels of atmospheric carbon dioxide (CO_2) had to be much higher than they are today. Carbon dioxide is not as strong a greenhouse gas as water vapor or methane and the increase in greenhouse warming caused by its presence becomes smaller as its atmospheric concentration rises. A small excess (over contemporary levels) could not have saved our planet from freezing during the faint-Sun era. Nevertheless, it was widely assumed that sufficient levels should have been around back then, so the hypothesis appeared a reasonable one.

It was also attractive because it could be made to fit easily into the so-called Gaia model; the hypothesis that life modifies its own environment in such a way as to make it more beneficial to itself. The point here is that the later appearance of photosynthetic organisms would have reduced the quantity of carbon dioxide in the atmosphere and reduced the greenhouse in sync with the increasing amount of solar radiation, in this way retaining a benevolent climate suited to the further multiplication of life.

However, a recent analysis of some of the world's oldest rocks by Professor Dennis Bird and his colleagues at the University of Stanford has dealt this neat hypothesis a serious blow. The rocks in question were recovered from Greenland and date back some 3.8 billion years. Because rocks react with their environment, their careful analysis can tell us much about what the planet was like at the time of their formation and, just as the zircon crystals from Western Australia changed our picture of late Hadean times, so these rocks from the far North seriously altered our opinion about Earth's atmosphere circa 3.8 billion years ago. In a paper published in April 2010, Minik Rosing, Norman Sleep, and Bird et al. demonstrate that the rocks, soon after their formation, had been exposed to levels of carbon dioxide very similar to those of today. The CO_2-rich atmosphere proposed to account for the hypothetical early greenhouse simply did not exist!

Moreover, the type of (microscopic) organisms that existed back then consumed carbon dioxide and hydrogen, but were quite sensitive to the concentration of these gases in the air to which they were exposed. In fact, the maximum levels of carbon dioxide found tolerable by these early organisms fell far short of that necessary to maintain a greenhouse effect sufficiently powerful to keep the planet from freezing. That is not to say that a greenhouse effect was absent altogether. On the contrary, greenhouse warming occurred then as it does today, but if the present greenhouse is required to keep the planet comfortably warm during this "hot Sun" epoch, it is obvious that a similar effect during those "cool Sun" days had to be inadequate.

Well, if the early Earth's "warm blanket" was not a greenhouse effect, what was it?

The answer, it seems, lies with the *albedo* or reflectivity of the early Earth. The albedo of an object is the measure of the proportion of incident light that it reflects back into space. The greater portion of light reflected, the higher the albedo; the less light reflected, the lower the albedo. But as we were told in secondary school physics, energy can neither be created nor destroyed (an oversimplified statement, but I did say "*secondary school* physics"!), so if more light is absorbed by a low-albedo object, the energy of that light has to go somewhere. It is, of course, transformed into heat. If two balls of the same material, but one painted white and the other black, are placed for an equal time in the sunshine, which one will be warmest when picked up? The black one, every time.

Climate modeler and co-author of the 2010 paper, Christian Bjerrum of the University of Copenhagen, found that the albedo of early Earth could be sufficiently lowered if the planet was then largely covered by ocean. In fact, at the dates being considered, there was very little dry land anywhere on Earth. Apart from a few islands, Earth was essentially "Ocean" and, as water absorbs more incident light (is "darker") than dry land, the average albedo of the planet must indeed have been considerably lower than it is today. Moreover, there would have been fewer clouds at those times; at least, fewer clouds of the type that significantly increase Earth's albedo. Clouds reflect some light back into space and therefore inevitably add to the overall albedo, but the clouds which do this most efficiently are the ones composed of an abundance of very small droplets. These clouds are formed when many tiny particles, capable of acting as nuclei for cloud-droplet formation, are present. Particles suitable for playing this role are abundant in the contemporary atmosphere of the planet, however most of them have a biological origin, generated by the vegetation that now abounds all over the Earth's surface. But 3.8 billion years ago, there was no vegetation. Nuclei for cloud formation were restricted to a little dust from the rare islands of exposed rocks and cosmic dust from meteors. Clouds were comprised mainly of larger and more transparent drops that were largely transparent to incident solar energy and did little to raise the albedo of the planet.

In view of this, one might say that early Earth was kept warm, not by a blanket but by a hot water bottle!

It is interesting to think what might have happened had water been less available on the early Earth. Even if the ratio of land to ocean had been similar to that of the present day, the land masses would have been without the darkening effect of soils and vegetation and the average albedo of Earth would have been too high to counter the cool Sun. As water froze and ice built up on the exposed land, the albedo would have continued to climb until the entire planet turned into one great ice ball. Probably, that is how it would have remained to this very day. On the other hand, as already hinted in the previous chapter, had the Earth remained covered by ocean – had significant land masses never emerged – the gradual increase in solar radiation and continuing low terrestrial albedo would presumably have resulted in increasing evaporation leading in turn to increasing water vapor content in the atmosphere. Water vapor is quite a strong greenhouse gas, so this would have raised the temperature still higher and may have precipitated a runaway global warming, heralding catastrophic consequences. The immediate reaction is to invoke increased cloudiness as a counter to this, but we should recall Bjerrum argument concerning clouds that form in the absence of very small (mostly biogenic) particles. Clouds of the early Earth may simply not have had the "right stuff" to reverse a runaway watery greenhouse.

Freezing Times on Planet Earth

Some 1.3 billion years ago or thereabouts, the steadily brightening Sun was shining enough heat on Earth to keep water liquid even assuming albedo and greenhouse conditions similar to those of the present day. Fortunately for us – and for the planet's biodiversity in general – increasing land area, the right sort of clouds and a general rising in the Earth's albedo marvelously balanced out the Sun's increasing output.

It would be incorrect however to think of these ancient epochs as ages of continuing climatic consistency. Quite the contrary. During the period between about 2.2 billion years ago and 635 million years ago, Earth experienced at least three prolonged periods of intense global cooling. It is possible that there was at least one even earlier global cold snap, but the cooling episodes of 2.2 billion years, 710 million years and 635 million years ago appear to be firmly established, although as we shall see shortly, there is some dispute as to just how severe they were. These periods were like ice ages on steroids. The picture which most scientists studying the climate of ancient Earth put forward was that of a totally frozen planet completely covered from pole to pole by continuous sheets of thick ice. Neither were these ice periods short lived. On the contrary, Earth remained encased in ice for millions of years at a time; the most recent of the super glaciations (the one 635 million years ago) is estimated to have lasted for between 6 and 12 million years. These times of intense cold have been graphically termed the *Snowball Earth* periods.

Not everyone entirely agrees with this picture however. Whilst there is general agreement that these cold epochs did happen and that they were of long duration, there *is* disagreement as to their intensity. Not all researchers are convinced that equatorial regions were completely under ice or that there was no liquid water anywhere on the planet's surface. Supporters of this milder picture have coined the term *Slushball Earth;* a somewhat less elegant-sounding, but (in their opinion) more accurate description of these early cold periods.

Whether the Earth was a snowball or a slushball, something had clearly gone wrong with its blanket! And it "went wrong" in a big way on at least three occasions.

It has been noted that the early period (ca. 2.2 billion years ago) corresponded more or less with the rise of photosynthetic organisms and the consequent rise of free atmospheric oxygen. This appeared to fit nicely with the interpretation of the earlier global warming as a carbon dioxide induced greenhouse effect. It seemed entirely reasonable to speculate that with the advent of photosynthetic organisms, the content of atmospheric CO_2 would fall as these organisms consumed it, causing the greenhouse to, so to speak, leak. Even though the Sun was shedding more heat 2.2 billion years ago than it had been around 3.8 billion years past, it was still cooler than it is today and the breaking of the greenhouse effectively meant that Earth shed the blanket that had kept it warm for the previous 2 billion years or thereabouts.

However, if the very early Earth was warmed, not principally by a greenhouse effect but by a low albedo, the role of photosynthetic organisms and the rise of atmospheric oxygen have presumably been greatly overestimated. Contrary to what promoters of the Gaia hypothesis may have hoped, the emergence of this new branch of life may have had less to do with the great chill than was previously thought.

The cause of these cold periods remains a subject of controversy. If fluctuating levels greenhouse gases were not responsible, maybe the cause is to be found in some external influence such as encounters between Earth and clouds of cosmic dust. That too has been put forward as a possible explanation. In the previous chapter, we mentioned that the Solar System appears to be near the galactic corotation radius and moves pretty much in sync with the propagation of the density waves defining the Galaxy's spiral arms. Although this works to spare the Sun (and Earth of course) from frequent crossings of the spiral arms, the Sun was probably not close to the corotation radius in its youth and crossings may have been more frequent then. This may be suggested as a possible explanation for at least the earlier cold periods, although it must be said that there is no real evidence for it and the suggestion has not gained wide currency amongst scientists.

Even though the cold periods themselves were pretty catastrophic, that does not necessarily imply that their cause was some major trauma. All that is really required is some trigger capable of setting off a period of global cooling. Once ice begins to build, the planet's albedo increases and temperature falls. This induces more ice and a further temperature decline, which in turn means more ice … Eventually, a low-temperature equilibrium is reached that persists until something occurs that is capable of reversing the trend. The wait may be a very long one indeed.

Whatever was responsible for the Snowball (or Slushball) periods, they (mercifully!) have not recurred during the last 500 million years. That is certainly not to say that there have not been cold periods and ice ages, the most severe and extensive being the Permo-Carboniferous glaciation between 310 and 285 million years ago. But nothing during this time equaled the earlier cold interludes.

It is interesting to note that the great proliferation of life known as the Cambrian explosion began about 100 million years after the most recent Snowball period. The first genuine land plants are thought to have started appearing about 470 million years ago (bacterial mats probably existed in marshy land regions earlier than this, but these were not strictly speaking *land* organisms) and the first land animals around 430 million years ago; some 200 million years after the last great cold epoch. Keeping the Stanford research in mind, it might be suggested that the colonization of the landmasses – first by plants and later by animals – probably lowered the albedo of these land areas by coating the "bright" rocks with "dark" vegetation and transforming bare rock into proper soil. On the other hand, more life also meant more biologically generated cloud droplet nuclei leading to an increase in the types of reflective cloud that tend to cool the lower atmosphere. Maybe the increased heat of the Sun, in concert with the lower albedo of landmasses, was sufficient to tip the balance in a generally warmer direction for most of the past half billion years.

Whatever the truth or otherwise of this, myriad other influences also come to bear on the climate of our planet. Some of these are clearly cyclic as we shall now see.

Cycles of Climate Change

Interned during the First World War, Serbian civil engineer and mathematician Milutin Milankovitch did not have much opportunity to do anything except think. Fortunately, he was pretty good at thinking, and his contemplations during those long days bore some very interesting fruit. What Milankovitch was able to work out during the time of his internment was a mathematical model showing that variations in the eccentricity of Earth's orbit, angle of the planet's axial tilt and orbital precession altered the patterns of climate. These cyclical changes in our climate have been given the name of *Milankovitch Cycles* in his honor.

These cycles mesh with one another in ways that are quite complex, but let's look at them one at a time to form some idea of what is going on here.

Taken in no particular order, let us look first at the cycle of precession of Earth's axis. This refers to the change over time in the direction of the planet's axis of rotation with respect to the "fixed" stars acting as a frame of reference. What part of the sky its axis is pointing toward, in effect. The Earth acts as a gyroscope, completing one full cycle in a period of approximately 26,000 years. After this period of time, our planet's axis points to the same stars as it did 26,000 years earlier. The cause of this gyroscopic motion has to do with the shape of the planet (i.e. an oblate spheroid instead of a perfect sphere) together with the tidal pull of both Sun and Moon.

During those phases of the cycle when the axis points more or less toward the Sun at the time Earth is at its perihelion (the point in its orbit nearest the Sun) one hemisphere will experience a greater difference between the seasons, other things being equal. On the other hand, during the parts of the cycle when perihelion and aphelion (the planet's furthest distance from the Sun) happen close to the equinoxes, both Northern and Southern Hemisphere seasons will be similar.

At the present time, perihelion occurs during the southern summer, a little over 2 weeks after the (southern) summer solstice. That should make the southern summer hotter than the northern and the winter correspondingly colder. The Earth is receiving about 6.8% more solar radiation during southern high summer than the Northern Hemisphere at the corresponding season, but because the Earth is closer to the Sun during the southern summer, it is also traveling faster in its orbit (according to Kepler's Law) causing the southern summer to be a little shorter than its northern counterpart.

Moreover, at the present epoch of its history, most of the Earth's landmass is concentrated in the Northern Hemisphere. The Southern Hemisphere is, by comparison, largely ocean. This more maritime environment modifies the climatic difference between the hemispheres. Large bodies of water have the effect of cooling the summer months and warming the winter and this effect takes the edge of severity off the southern seasonal variations. Climatically, Earth would be a rather more extreme place during the current epoch if most of the land was in the Southern Hemisphere.

In addition to its gyroscopic motion, the planet's axis also varies in its degree of tilt or obliquity. The obliquity is currently at 23.44° from the orbital plane (using the latter as the plane of reference) but slowly varies from 22.1 to 24.5 and back again over a time scale of 41,000 years. We are at present in the decreasing phase of the cycle, with minimum value set for around the year 10,000 AD. As obliquity steadily declines in future millennia, the (very!) long-range weather forecast is for an "evening out" of the climate, with warmer winters and cooler summer ... other things – as usual – being equal.

Computer models incorporating more extreme variations in obliquity indicate that variations in climate would then become so severe as to threaten the very existence of civilization or even of complex warm-blooded life itself. We will return to this, and to the mechanism preventing such variations from happening, later in this chapter.

Thirdly, the shape of Earth's orbit goes through cyclic variations. If ours was the only planet orbiting the Sun, its orbit would remain pretty much constant for great periods of time, excepting a close encounter between the Sun and a passing star or something equally as dramatic. However, ours is far from being the only planet in the solar family, and the gravitational perturbations of these – especially the gas giants Jupiter and Saturn – cause the eccentricity of Earth's orbit to vary from a nearly circular 0.005 to a more distinctly elliptical 0.058. This is still only slightly elliptical as elliptical orbits go (that of Mars, for instance, is 0.0934) but the variation is enough to make its presence felt. The present value of the orbit's eccentricity is 0.017. The variations in our planet's orbit do not trace a simple cycle, but the major variation happens on a period of 413,000 years and covers an eccentricity range of plus or minus 0.012. Other variations occur with periods between 95,000 years and 125,000 years, with a beat period of some 400,000 years, loosely combining to form a cycle having a length of 100,000 years and a variation of eccentricities ranging from 0.03 below the mean to 0.02 above it.

Today, the very moderate eccentricity of earth's orbit results in the planet receiving 6.8% more solar heat at perihelion than at aphelion. The climatic effects of this variation, as previously mentioned, is largely ameliorated by the timing of perihelion with the southern summer and aphelion with southern winter. When the orbit is at its most eccentric, the difference rises to about 23%. We should not expect conditions to be so benign when Earth's orbital eccentricity is at maximum and perihelion coincides with, or occurs several weeks after, the northern summer solstice (assuming a landmass distribution similar to that of today).

Fourthly, it is recognized that the Earth's elliptical orbit also precesses, or slowly rotates, in space. This is known as apsidal precession and, more precisely, is defined as the gradual rotation of the line joining the apsides (i.e. the perihelion and aphelion points) of the orbit. Like the variations in the orbit's eccentricity, apsidal precession is also primarily due to gravitational effects of the giant planets Jupiter and Saturn. Because the Earth's orbital precession is in the same sense as the gyroscopic motion of its rotational axis, the date of perihelion repeats on a cycle of about 21,636 years rather than nearly 25,772 had the gyroscopic motion been omitted. The dates of perihelion and aphelion advance on an average of 1 day for each 58 years.

Fifthly, the inclination of the orbit also changes, drifting up and down with a cycle of about 70,000 years. It has also been noted that the Earth's orbit moves relative to the invariable plane of the Solar System (effectively, the orbital plane of the largest solar planet, Jupiter). The inclination of Earth's orbit varies with respect to the invariable plane on a cycle of 100,000 years. Milankovitch himself did not study either of these cycles, but it is a curious coincidence that the 100,000-year cycle matches the 100,000-year variation of orbital eccentricity. Even more curiously, records of past climatic changes over the last million years reveal a clear pattern of ice ages recurring on a cycle of ... 100,000 years! Could both of these be *mere* coincidences?

That is an open question.

With respect to the apparent coincidence of variations in orbital inclination and ice ages, the existence of a concentration of cosmic dust along the plane of the Solar System has been proposed and the hypothesis put forward that during epochs of very low orbital inclination, Earth remains essentially confined to this dusty torus throughout its journey around the Sun. The relatively high amounts of dust sweeping down into our atmosphere might be sufficient to raise atmospheric opacity as well as acting as nuclei for increased cloud formation. The global cooling resulting from this may be enough to act as a trigger for ice age development. Evidence for the existence of a dust torus has been cited in the increase of faint meteors detected by radar around January 9 and July 9; the current dates on which Earth crosses the invariant plane. The incidence of noctilucent clouds (see Chap. 6), possibly triggered by an influx of meteor dust, may also increase around these dates.

On the other hand, other researchers argue that the timeline of one million years is not long enough to establish a cycle of 100,000 years with statistical significance. Perhaps it is simple coincidence after all!

Much more secure, however, is the association between the intensity of Earth's seasons and the variations in the planet's axial obliquity, orbital eccentricity variations and changes in the relationship between solstices and dates of perihelion/aphelion. Though the *direct* effect of this combination of factors might not be large, they act as triggers for an overall change in Earth's climate through feedback mechanisms. Greater snowfall during Northern Hemisphere winters, coupled with cooler summers allowing more ice and snow to remain without melting, raise the albedo of northern continental landmasses. By reflecting more solar radiation back into space, this higher albedo causes temperatures to drop still further, meaning more snow, more ice, and a further increase in albedo, starting the cycle all over again.

Has Our Solar System Environment Modified Climate Change?

We noted earlier that computer models incorporating larger variations in obliquity than actually experienced by our planet resulted in climatic changes that, if they occurred on the real Earth, would have spelt the end of civilization and probably the extinction (or more likely the non-appearance in the first place) of humans and other complex warm-blooded creatures. An exaggerated counterpart of the feedback mechanism that most likely triggers our real ice ages plays a large role in these hypothetical cases. But this raises an interesting question. We know that Mars is a planet that really does experience enormous swings in axial obliquity and it is widely believed that part (although not all) of the reason for lack of complex warm-blooded creatures there is due to this fact. The question that raises its head at this point is why Earth does not go through similar large variations in obliquity. This in turn raises a further issue, viz. whether our presence here is dependent upon the small range in obliquity variation experienced by Earth or whether (Mars notwithstanding) life has simply adapted to terrestrial conditions and may have equally adapted to different circumstances elsewhere.

This issues goes to the heart of the so-called Anthropic Principle; the principle that – simply stated – we observe the apparent fine tuning of the environment because if it did not possess these fine-tuned properties, we and all conscious life (what we termed "androde" life in the previous chapter) would not be around to observe it. This principle takes various forms. For some, the fine tuning is a matter of pure coincidence, while others (for example, British cosmologist Sir Fred Hoyle) argued for the presence of a designing intelligence. The counter hypothesis is to assume that life is almost infinitely adaptable and that even organisms having the complexity of androdes are capable of evolving in extreme environments through natural selection. Those taking this line tend to dismiss the Anthropic Principle in whatever form it is presented, as being untestable and therefore "not a true scientific hypothesis." Proponents of the Anthropic Principle in their turn point to the lack of advanced life on the other planets of the Solar System and the low levels of biodiversity in the harsher environments of our home planet. If life exists on Mars or Titan or Europa, etc. nobody expects it to include verdant diminutive humanoids ... or even a richly biodiverse ecosystem. So clearly, there are limits to the adaptability of, at the very least, advanced life. If simple life is found on Mars, say, far from supporting the view that once life starts it will keep on evolving in any environment, it will actually demonstrate the point raised in the previous chapter; life has its limits and these limits tighten as increasingly complex organisms are considered.

In a 2011 paper published in *Astrobiology*, University of London scientist Dave Waltham tackled this issue by analyzing two apparent instance of fine tuning, viz. the axial precession rate of Earth and eccentricity variations of Earth's orbit.

The former is controlled by the planet's rate of spin and by the size of tidal forces exerted by both Sun and Moon on its equatorial bulge. The size of the Moon has been found to play a major role in the Earth's axial stability and the properties of the Moon depend upon the exact circumstances of the giant impact between Earth and another planetary body some 4.5 billion years ago. Waltham found that if the Moon was a little smaller or a little larger, the axial stability of Earth would be significantly less than it is in reality, resulting in a more rapid rate of precession. Indeed, the actual size of Earth's Moon lies within the narrow limits permitting the great degree of axial stability experienced by our planet. Outside these limits, this Milankovitch cycle would be much more extreme. Not only would cycles of climate change be more severe, but they would also be more frequent.

By running computer simulations of the formation of Earth/ Moon systems, Waltham found that the probability of our finding ourselves, purely by chance, in a system capable of giving the slow Milankovitch cycle experienced by the real Earth is as low as 0.77%!

He then examined the effects that changing the locations of the other Solar System planets would have on the eccentricity variations of Earth's orbit, and once again found that altering these from their true positions would destabilize the situation and result in faster Milankovitch cycles of eccentricity variation. The probability that we find ourselves by pure chance in a system allowing such mild and slow variations in orbital eccentricity is calculated as just 3.2%.

Lastly, Waltham examines the overall structure of the Solar System from the point of view of the near-commensurability in the orbital periods of the major planets. An instance of this is the Jupiter and Saturn pairing; the time taken for Jupiter to orbit the Sun five times is almost the same as that needed by Saturn to go around its own orbit twice (59.3 years as against 58.92 years). This phenomenon suggests that the positions of the different planets in a system cannot be taken as independent variables. Something "orders" the structure of a system of planets as a whole. Work performed during the last few years suggests that processes in the early Solar System, and probably in all planetary systems during their formative epochs, incline the evolution of these systems toward near-commensurabilities in orbital periods. These processes include such things as gas drag and planet-planet scattering.

Modeling for 10,000 randomly chosen hypothetical planetary systems, Waltham found that just 386 turned out to have the structure of our own Solar System, i.e. systems where slow cycles of eccentricity in the orbit of a planet like Earth are permitted. From the consideration of the various values of near-commensurabilities, the chances of finding ourselves in such a system therefore comes out at 3.9%.

Considering all three of these tests, Waltham concludes that "The probability that the results from all three of these tests are purely coincidental is less than one in 10^5."

He draws the following four conclusions from this study:

1. Anthropic proposals can be tested by comparing Earth's properties to those of synthetic populations of Earth-like worlds generated with computer modeling.
2. Investigation of Earth's Milankovitch cycles shows that these are relatively slow, and this may indicate anthropic selection for slow climate change.
3. The Gaia hypothesis may have confused cause and effect, that is, climate stability might be a precondition for a complex biosphere rather than climate stability being the consequence of a complex biosphere.
4. Planets with Earth-like levels of biodiversity may therefore be rare in the Universe.

Slow cycles of climate change appear, therefore, to be very important in maintaining our planet as the green home that we know and love. Moreover, it seems that the Solar System environment has played a vital role in ensuring that the cycles were, indeed, slow. But the cycles are not the whole story. The geological record tells of events which can punctuate the regular cycles and bring about sudden changes that are truly catastrophic. To the most famous of these, we shall now turn.

The Great K-T Catastrophe

About 65 million years ago, a dramatic change occurred on Planet Earth. The Cretaceous era – the time when dinosaurs roamed the Earth – suddenly gave way to the Tertiary. Some major climatic and ecological upheaval took place resulting in mass extinctions and the emergence of a new ecosystem. Most notably in the popular mind, the dinosaurs vanished from the face of the globe, paving the way for the flourishing of mammalian species and, eventually, humankind.

The cause of this sudden catastrophe has been the subject of much speculation. Was an outbreak of volcanism responsible? Did a supernova explosion relatively close to the Solar System bath the planet in lethal doses of radiation? Did a large astronomical object (an asteroid or comet) strike the Earth? Or was the cause something that nobody has thought of as yet?

An important clue was unearthed in the late 1970s when Luis and Walter Alvarez of the University of California at Berkeley discovered an abnormally high concentration of iridium in the thin layer of clay marking the boundary between the Cretaceous and Tertiary eras (the so-called K-T boundary). Levels of iridium within this layer reached six parts per billion. This may not seem to be high, but compared to an average for the Earth's crust of just 0.4 parts per billion it is a very significant concentration. Importantly, meteorites have far larger iridium concentrations – up to around 470 parts per billion – so the relatively high concentrations in the K-T layer strongly implied a large meteorite impact followed by a spread of meteoric dust around the globe during the following years. Mixed with ordinary terrestrial dust, this settled into the discovered clay layer, thereby explaining the latter's elevated iridium levels.

Fig. 2.1. K-T boundary layer exposed by erosion on rocky hillside, Badlands near Drumheller, Alberta (Credit: G. Larson)

At the time of this discovery, no crater was known dating from the time of the K-T boundary. A number of suggestions were put forward, including Fred Whipple's hypothesis that the impact may have been so great as to penetrate Earth's crust and leave, not a crater, but an up welling of fresh magma. Whipple suggested that instead of seeking a crater, we should look for a relatively fresh region of intense volcanism and suggested that Iceland might be a good possibility. This suggestion, though interesting and refreshingly "outside the box" nevertheless fell by the wayside for lack of supporting evidence.

A far more promising candidate emerged in the form of the Chicxulub crater, an "astrobleme" or ancient impact crater buried beneath the Yucatan Peninsula in Mexico. The first major hint of something interesting in this area came as far back as 1951 when exploratory borings in search of oil hit what at the time was thought to be a dome of solidified lava some 4,200 ft (1.3 km) beneath the surface. Lava domes are not a feature of the area, but nothing further was found until the late 1970s when geophysicist Glen Penfield, in the employ of Mexican state-owned oil company Pemex, discovered a strange geological feature buried beneath the

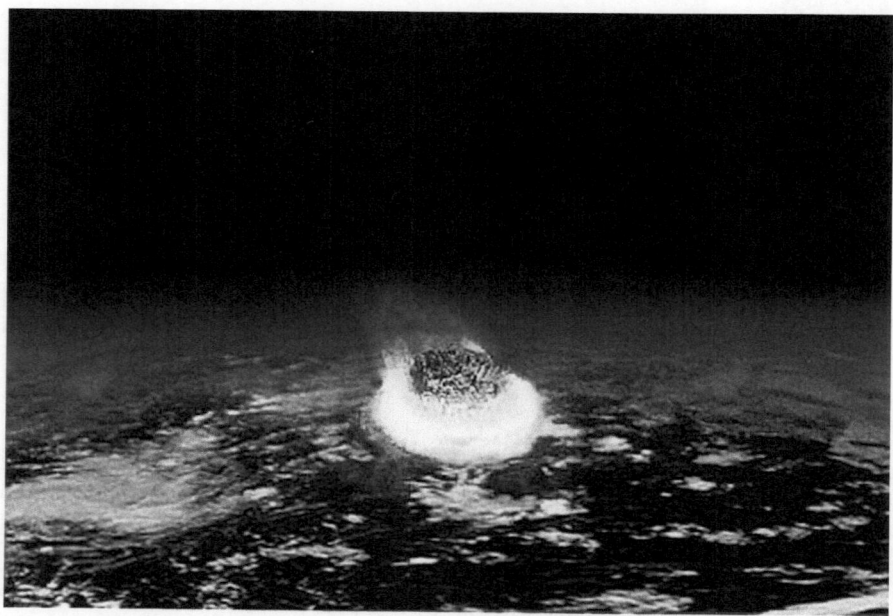

Fig. 2.2. Artist's impression of giant impact event (Credit: NASA)

peninsula. This find, it might be noted, took place about the same time as the Alvarez discovery but news of it did not spread very far for several years and had certainly not reached Berkeley at the time of the Alvarez research.

Then, in 1990, reporter Carlos Byars informed Alan Hildebrand of the University of Arizona of Penfield's discovery. As Hildebrand had previously sought a candidate crater and was aware of suspicious deposits in the Caribbean region, this piece of news must have been music to his ears!

In the April of 1990, Hildebrand contacted Penfield and the two quickly secured two Pemex drill samples stored in New Orleans. Both samples revealed evidence of shock-metamorphic alteration of the type expected for material affected by a violent meteoritic impact.

Later in that decade, examination of satellite images revealed the diameter of the ancient crater to be some 190 miles (300 km) across! The blast needed to gouge out a hole of that size is estimated to have been of the order of 100 million megatons or around two million times more powerful than the largest thermonuclear device ever exploded by mankind. The object that caused this mighty blast is estimated to have been around 6 miles (10 km) in diameter. Such a body would have been quite capable of blowing enough meteoric dust into the air to account for the iridium-rich layer discovered by Luis and Walter Alvarez.

It did much more than that of course!

There is evidence that the region where the body impacted was covered by deep water at that distant time. The immediate result would have been mega tsunamis thousands of feet high – literally mountains of water – sweeping across ocean and land, taking all in their path. Clouds of super-heated steam, dust and ash spread from the crater, billowing high into the atmosphere. Shock waves radiating outward from the impact site may have triggered earthquakes and volcanic eruptions in other places as Earth answered this cosmic assault with plutonic violence of its own. This would have had the effect of adding sulfate aerosols to the atmosphere, over and above the large quantities of particles already belched out from the blast itself. These sulfates would have had two detrimental effects spreading far beyond the regions immediately affected by blast and heat. Sulfate aerosols spreading through the upper atmosphere would have triggered a very significant global cooling, while those in the lower air capable of being washed out by rain would have greatly added to the acidity of the rainwater, with all the associated detrimental effects on vegetation.

But before either effect became widespread, wildfires swept large regions of the planet, also releasing huge volumes of smoke and ash. Carbon dioxide released from carbonate rocks in the impact zone may have spawned a greenhouse warming initially, but later the sun screening effect of high-altitude sulfate aerosols, smoke and dust, would have precipitated a world-wide cooling and a significant darkening of the daylight hours that may have lasted for several decades. Photosynthesis was interrupted, leading to global die-back of many plant species. This directly affected the survival prospects of herbivores and indirectly that of the carnivores which preyed upon them. This reduction in plant life may have been partially offset by the increased carbon dioxide in the atmosphere, but the trauma to the entire ecosystem inevitably resulted in numerous extinctions, especially of species which were already struggling to survive for one reason or another.

But did the ecological disruption caused by the impact really kill the dinosaurs? Or was it just one of a number of factors that led to – and probably hastened – their demise?

Popular discussions have given the impression that this was *the* dinosaur's downfall – full stop. But many paleontologists remain unconvinced that it tells the whole story. Some have argued that the large dinosaurs had simply grown too big to be as efficient as a changing world demanded. Others deny that the dinosaurs as a species became extinct at all. They are still with us today, albeit (fortunately!) smaller and sporting wings and feathers. We hear them twittering in the trees on most mornings!

This latter thesis has also become more complex in recent years with the discovery of fossilized bird skeletons dating back millions of years *before* the K-T boundary. Apparently birds not unlike those of the present time co-existed with the dinosaurs, in the later stages of the latter's reign at least. Some of these ancient birds became extinct; others did not. For some reason not yet clear, birds which could broadly be called "duck-like" were the survivors.

Furthermore, as paleontologist Robert Bakker points out, frogs were also contemporary with the dinosaurs and even depended upon a similar environment. If the effects of the Chicxulub impact were as deadly to species as traditionally portrayed, how did frogs manage to cross the K-T boundary while dinosaurs did not?

To pursue this any further would venture beyond both the subject of this book and the competence of its author. Suffice to say that the subject is not as simple as often presented and the last word on the dinosaurs' demise is unlikely to have been written.

Yet, what *is* no longer in doubt is the occurrence of an enormous collision about 65 million years ago and a massive climatological and ecological shock resulting from this. Whether the meteorite was the total cause or not, its arrival saw the disappearance of one era of life on this planet and the birth of another. Yet, however serious and devastating the event was, life survived and the ecosystem restored itself in what was, geologically speaking, but a moment of time. If the thought of an asteroid collision engenders fear, this corresponding fact should equally inspire hope in the long-term survival of our blue world.

Whence Came the Asteroid?

The size of the body that struck Earth 65 million years ago was greater than any asteroid capable of making close approaches to our world today. Moreover, it would appear that the Earth was hit by a number of smaller objects (in the order of tens of meters across) around the same time as the Chicxulub meteorite. At least two smaller craters are known from that era (Silverpit in the North Sea and Boltysh in Ukraine) with a third (Shiva) of disputed impact origin. Then, going back further into Solar System history, the Moon was struck by an asteroid somewhat smaller than the Chicxulub object around 108 million years ago, forming the conspicuous Tycho crater. (Incidentally, this event was undoubtedly followed by remarkable meteor storms on Earth as our planet swept through ejected lunar debris. Some of the debris pieces may well have been large enough to land as meteorites.)

It seems that the last 60 million years have been pretty quiet in the Earth-Moon neighborhood compared with the previous period of equal duration (that is certainly not a complaint, please note!). It would appear that there was a heightened influx of asteroidal bodies into the inner planetary system at that earlier time.

Examination of meteoritic particles, and a fossil meteorite, from the time of the Chicxulub impact indicate that the body had the composition of a CM carbonaceous chondrite; an interesting class of meteorite that we already met in the previous chapter. Knowing the type of meteorite that formed the Chicxulub crater might give some indication of that body's origin.

In 2007, a joint U.S. and Czech team of scientists including W. Bottke, D. Vokrouhlicky and D. Nesvorny, studied evidence of an event believed to have taken place within the inner asteroid belt around 160 million years ago.

Many of the asteroids of the Main Belt exist in groups or "families" sharing somewhat similar orbits, a fact discovered as long ago as 1918 thanks to the work of Kiyotsugu Hirayama. These "Hirayama families", as they are known, arise from the collision of two relatively large asteroids. It was one such family of asteroids that caught the interest of Bottke et al. namely, the so-called

Fig. 2.3. Asteroid 298 Baptistina. Magnitudes of nearby stars marked for comparison (Credit: Kevin Heider)

Baptistina family, so named for its brightest member, the asteroid 298 Baptistina. According to this team, an asteroid some 106 miles (170 km) in diameter, was struck by a second body of about 37 miles (60 km) diameter approximately 160 million years ago, creating this family of asteroids.

Over time, the Baptistina asteroids slowly spread out through nearby space, with a number of them eventually drifting into a region where the gravitational influence of Jupiter deflected some into orbits that crossed the path of the Earth and Moon. The team estimated that as many as 20% of the asteroid fragments may have ended up in orbits that showed little similarity to those of the core family members. About two percent of these errant objects are estimated to have struck the Earth.

The time-line appears to agree quite well with the impacts on Earth and Moon some 60–100 million years later, including the lunar Tycho impact and the Chicxulub meteorite. Moreover, a study of the visual spectrum of Baptistina and another of the brighter members of its family in 2004 indicated that these objects are C-Type asteroids, having a composition similar to that of the Murchison meteorite. This would nicely fit with the proposed Baptistina origin of the Chicxulub meteorite.

However, all did not run as smoothly as might have been hoped. A later study by a team of scientists (including V. Reddy, as well as Bottke and Nesvorny) of the near infrared spectrum of Baptistina yielded results that were at variance with the earlier one. Reddy and team found that the spectrum of this asteroid showed little similarity with that of a typical C-type. Indeed, even the visual spectrum suggested a weak feature near 0.9 μm that is not present in C-type objects, but *is* found in the spectra of S-type asteroids; stony, siliceous bodies similar to ordinary, non-carbonaceous, chondritic meteorites. The new study confirmed this and also yielded an approximate estimate of the object's albedo (reflectivity) of between 0.09 and 0.16, comparable to the average S-type of 0.1 or higher. Typical C-types, on the other hand, are in the region of 0.05 or less.

Yet, in spite of these S-type indications, the spectrum did not fall into this class either. It has been listed in the rather broad class of X-type; a little like the "Any Other Variety" at a dog show. But the true nature of the asteroid is difficult to determine. Reddy et. al. suggest that the nearest meteorite analogues might be CO or CV chondrites (carbonaceous chondrites that have experienced greater metamorphic alteration than the CM types), or the even more highly metamorphosed carbonaceous meteorites known as ureilites or the non-carbonaceous pallasites. The latter may results from shock-induced alteration of rocks brought about in violent collisions between asteroids, whereas the ureilites have, since October 2008, become associated with asteroids of the F-type, a sub-species of the C-types that show a featureless or "flat" spectrum. That month, a tiny asteroid – or, more accurately, a large meteoroid – designated 2008 TC_3 was discovered on collision course with Earth. Fortunately, a spectrum of the object was obtained prior to its atmospheric entry and this identified it as being of the F-type. The object entered the atmosphere over the Sudan and meteoritic fragments subsequently recovered from the Nubian Desert were identified as ureilites … albeit with some peculiarities.

The best bet for Baptistina might be a composition similar to carbonaceous chondrites of the CO or CV type. If so, it might resemble the Allende meteorite more closely than Murchison or, probably, Chicxulub.

Nevertheless, Reddy and colleagues do not think that this apparent discrepancy necessarily exonerates Baptistina from all involvement with the K-T event. As the largest member of its family, this asteroid may comprise the core and innermost regions of the original body and would therefore have been more highly pressure cooked than the outer parts of the parent asteroid. Assuming the bulk of the original asteroid to have been composed of CM-like material, it should come as no surprise that its deeper parts became more highly metamorphosed under these conditions.

Moreover, it might be relevant to note that the Baptistina family overlaps a larger and older Hirayama family named for its brightest member, 8 Flora. The Flora family members are S-type. There is no debate about this, but because they occupy a well populated region of the Main Belt, the family includes a number of pseudo members, several of which are identified as gate crashers by means of their dissimilar classification. The parent asteroid of the Baptistina family can be regarded as having been one of these interlopers; one that gave birth to an entire family of interlopers. Most likely, the smaller body that struck it was one of the Flora members, and as such an S-type. If this is true, many Baptistina family members may share a mixture of C and S material, making clear classification difficult. Moreover, even today infrared images reveal a band of dust coinciding with the zone of the Flora asteroids, presumably raised from their surfaces by meteoroid strikes. Surely this stuff is also swept up by asteroids moving through this diffuse stream of dust and gravel. In short, it might be suggested that some of the features detected in the spectra of Baptistina reflect not so much the asteroid's bulk composition as the degree that detritus from the Flora family has accumulated on its surface. Might the 0.9-μm feature be explicable in this way?

In any case, early in 2010 a strange object was found that might add yet another piece to the jigsaw puzzle. Let us now take a closer look at this oddity.

The Case of the Decapitated Comet

On January 6, 2010, the Lincoln Observatory (LINEAR) sky survey discovered a nebulous streak, having the appearance of a comet tail that had lost its head. Or, to be more precise, like a comet's anti-tail that had lost its head, as it turned out that the apparent tail pointed sunward; the brightest portion being that furthest from the Sun.

Initial observations of this "headless horror" allowed a very rough preliminary orbit to be calculated suggesting a short-period ellipse typical of the comet population within the inner Solar System. As further observations were made, the orbit was drastically revised. It remained of short period, but the upgraded orbit resembled that of a Main-Belt asteroid rather than a typical short-period comet.

In one sense, that was not too surprising. Four Main-Belt comets (essentially asteroids which go through periods of cometary activity) had already been found and the narrow, bright, tails of these objects were not too unlike that of this new object; P/2010 A2 (LINEAR) as it was officially known.

There was one major difference however. All four previous discoveries are located in the outer Main Belt, where C-Type bodies form the overwhelming bulk of the population. Three of the four are associated with the Themis Hirayama family of C-Type

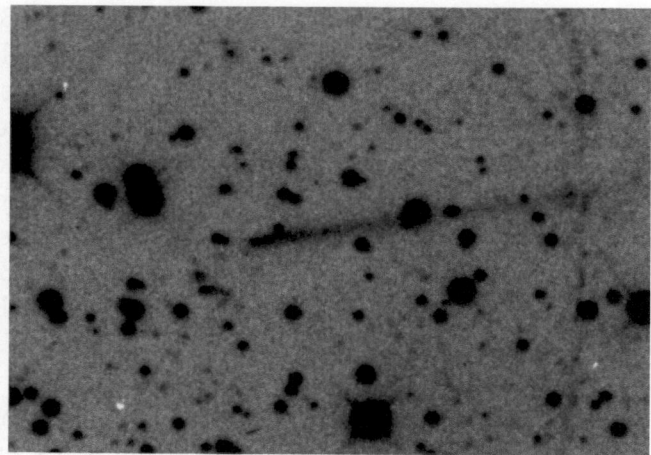

Fig. 2.4. Main Belt Comet P/2010 A2, January 12, 2010. Eight minute exposure using 24-in. telescope (Credit: Kevin Heider)

asteroids; two clearly within the family, while the third seems to be an escapee from this formation. It is not too surprising to find that at least some Themis-family objects retain enough ice to produce periods of cometary activity when, for instance, a portion of sub-surface ice is exposed after a meteorite strike or some such event. Indeed, both ice and a very comet-like mixture of organic materials have been detected on the surface of the family's namesake and main member – 24 Themis. Another Main Belt comet discovered a few months after P/2010 A2 also proved to be a denizen of the outer Belt.

But 2010 A2 does not lie in that region. Its orbit places it firmly within the inner Belt where waterless S-Type asteroids predominate. Indeed, from its orbital elements, it appears to be a member of the Flora family of asteroids which, we will recall, are of the S-Type. The Flora family is believed to be the home of L Chondrite meteorites, as well as the main source of siliceous cosmic dust particles collected in Earth's upper atmosphere. Both L Chondrites and siliceous dust show no evidence of having been in contact with water. Quite the contrary in fact, which agrees well with the accepted model for the formation of the Solar System. S-Type bodies are thought to have formed within the inner regions of the pre-solar nebula, where temperatures were too high for water and other volatiles to exist.

Assuming therefore, that 2010 A2 is a bona fide member of the Flora family and therefore an S-Type body, how can it retain the ice necessary for the observed cometary episode?

According to all the accepted wisdom on the subject, it cannot. Perhaps the tail was not really evidence of "true" cometary activity at all!

This was the line taken by most astronomers following publication of the object's true orbit. The widely held opinion interpreted the "tail" as a trail of debris released by a very recent collision between two previously unknown asteroids within the Flora family. The sunward-pointing orientation of the tail implied relatively large particles, consistent with the collision hypothesis, although not necessarily requiring it. Moreover, images from the Hubble Space Telescope in late January 2010 revealed a peculiar dust formation, at the terminus of the brighter portion of the tail, looking rather like the letter "X." This superficially suggested a violent – even explosive – event in the very recent past. The arms

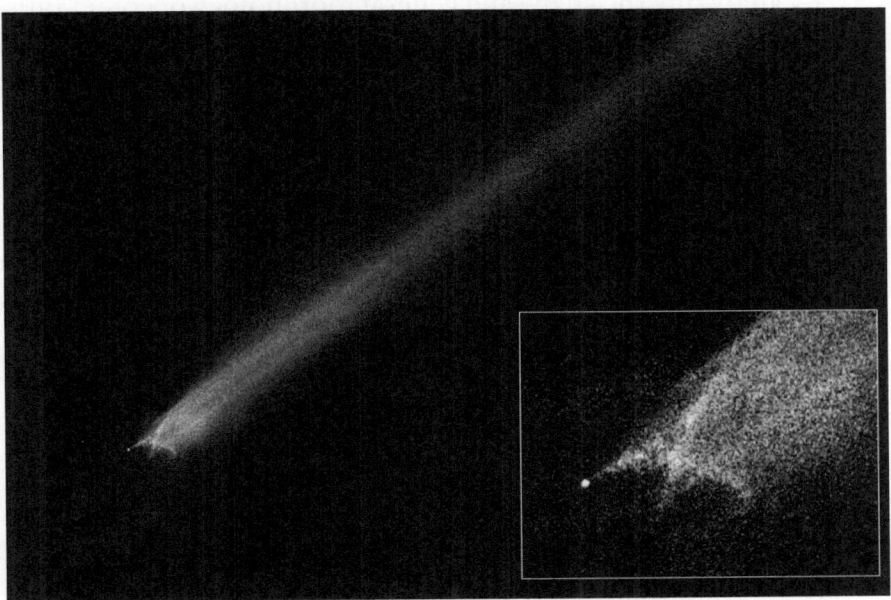

Fig. 2.5. Main Belt Comet P/2010 A2, February 3, 2010 as imaged by Hubble Space Telescope showing peculiar X-shaped feature and asteroidal fragment (Credit: NASA, ESA, D. Jewitt (UCLA))

of the "X" might have been formed by crumbling fragments thrown out from the asteroid crash. Moreover, a few days after the object's discovery, a very faint asteroid-like object was found just outside the tail, and well off centre from the tail's "axis." On the collision hypothesis, this "nucleus" was interpreted as the largest remnant of the shattered asteroids, kicked out and to the side of the debris train by the force of impact.

The asteroid collision scenario seems to explain the peculiarities in appearance of the new object, as well as removing the awkward issue of ice existing inside an S-Type asteroid. True, it does rely on a rather favorable timing; LINEAR observing the object shortly after the event, but as collisions between asteroids certainly occur from time to time, the co-incidence is not too farfetched.

Nevertheless, doubt was thrown on the collision hypothesis by Fernando Moreno and colleagues, who performed a detailed analysis of the dust tail based upon a series of images obtained with three telescopes at the *Observatorio del Roque de los Muchachos* on La Palma in Spain between January 14 and 23, 2010.

By analyzing the shape and general morphology of comet tails, it is possible to determine both the bulk physical properties of the particles themselves and the times they were released from the cometary nucleus. Applying this analysis to the tail of 2010 A2, Moreno et. al. confirmed the earlier conclusion that the particles comprising it were rather large by the standards of dust motes; up to approximately 1 cm in diameter in fact.

However, these researchers did not confirm the assumption that all of these particles were released in a single violent burst as expected from a collision of asteroids. On the contrary, their research suggested that particle release began quite mildly around the end of March 2009, increasing until June of that year, then steadying off for a while before declining until early December. This was also when the comet was at its closest point to the Sun, but as the orbit is nearly circular, there is probably little significance in that.

This sounds very much like a true cometary outburst. Moreover, the Moreno analysis also indicated a velocity of ejection for the particles that was entirely consistent with evaporating ice being the source of activity, again suggesting a "classical" cometary scenario. In short, the observed tail was pretty much what would be expected from a comet of about 200 m diameter at a distance of roughly twice that of the Earth from the Sun; a very good fit to the comet's actual estimated size and distance.

Because the comet was not discovered until after this hypothesized activity had ceased, any gas and fine dust would have already dispersed by then, leaving only a trail of heavy gravel. Hence the "headless" appearance at and following discovery.

The Moreno analysis also suggested that activity was probably confined to a single region in the high southern hemisphere of the nucleus. Contrary to the earlier suggestion that the nucleus was merely a fragment knocked out of the tail by the hypothetical collision, the Moreno model sees its off-centre position as a consequence of this lopsided emission of tail particles.

The process may indeed have been started by an impact, but (on Moreno's scenario) that was far from being the end of the story.

The Moreno position was, however, quickly challenged. Contrary to his conclusions, analysis of space-based observations by the *Rosetta OSIRIS* team indicates a sudden formation of the tail in a single burst occurring early in February 2009. In apparent support of this analysis, D. Jewitt, H. Weaver et al. find that their study of *Hubble* images also fit a sudden and brief burst either late in February or early in March 2009. Both of these analyses better fit the collision scenario than one of extended cometary activity.

Whichever scenario turns out to be correct however, it is interesting to note that the orbit of P/2010 A2 shows greater similarities to that of Baptistina than to Flora. Perhaps the comet is not an S-Type after all. It may indeed be related to icy objects such as "genuine" comets!

Although Baptistina is no longer considered a C-Type object, it may still be related to these bodies. If the Baptistina family truly did result from a collisional breakup of a large asteroid with C-Type outer layers and a more differentiated core (now remaining as Baptistina itself), at some depth beneath the surface of the progenitor, ice may have been stable. It could persist even today in some of its fragments. Maybe 2010 A2 is one of them.

If 2010 A2 is a member of the Baptistina family and if the relationship between the Chicxulub meteorite and Baptistina holds, this Main-Belt comet and that meteorite are related, however distantly!

By the way, it should be made clear that 2010 A2 is not about to mimic its possible distant cousin and hit the Earth, as some sensationalist rumors have said. Its orbit is stable and lies well beyond the Earth. At its closest possible approach to our planet, the comet's distance from us is equal to that of the Sun; not any cause for worry we will surely agree!

Earlier Extinction Events

The KT extinction, although the most recent and the best known in popular literature, is far from the only such event in Earth's history. An even greater extinction event marked the transition from the Permian to the Triassic period, about 245 million years ago, in which it is estimated that as many as 90% of species, 75% of terrestrial vertebrates, 50% of genera and 33% of marine mammals vanished.

In addition to these two major extinction events, there have also been somewhat less devastating ones which, according to paleontologists J. Sepkoski and D. Raup, happen at a regular rate of every 26 million years. This apparent periodicity led a number of astronomers to envisage a remarkable scenario. The Sun is not the single star that we generally believe it to be! Instead, it has a small, faint and remote companion orbiting in an elongated ellipse with a period of around 26 million years. At its furthest, this solar companion has little observable effect on the other members of the Sun's family but its closest approaches bring it within the Oort Cloud or vast system of comets hypothesized to surround the Solar System far beyond the orbit of Pluto. When this happens, the gravitational influence of this star knocks thousands of these objects from their orbits, some drifting off into interstellar space, but others plunging downwards toward the Sun and central planetary system. Some of these, according to this hypothesis, collide with the inner planets including Earth and it is these impacts that act as triggers for the ecological disruption leading to mass extinction events.

The hypothesis has a certain ring of excitement about it, but the star of the story is missing. There is simply no corroborating evidence for the existence of a stellar companion of the Sun. *Nemesis* – the Death Star – as it has come to be known, appears to be nothing more than an interesting hypothesis.

Nevertheless, there may be other reasons why Oort Cloud comets may shower down on us. Attention has already been drawn to the Sun's favorable galactic location close to the corotation circle where stellar orbital velocity matches the speed at which the density waves compressing the Galaxy's spiral arms are moving. Located outside the main spiral arms, this position is a healthy one as, while ever the relationship is maintained, dangerous spiral arm encounters are kept to a minimum. But the Sun was not always at this favored location. Maybe spiral arm encounters were more frequent once and the earlier extinction episodes at least may have resulted from encounters between the Oort Cloud and the giant molecular clouds strung along the Galaxy's spiral arms.

Once again, there is no compelling evidence to back this up and, although encounters with cosmic clouds and individual passing stars surely happen from time to time, further investigation of the supposed periodicity of extinction events failed to confirm Sepkoski's and Raup's results. The alleged 26 million-year cycle turned out to be nothing more substantial than an artifact of the way in which the data had been analyzed.

In addition to this, if our earlier discussion regarding the cause of the K-T event is correct, the Oort Cloud played no part in at least one mass extinction. The real culprit lay a lot closer to hand; within the inner asteroid belt. Moreover, the even more dramatic Permo-Triassic extinction appears to have coincided with widespread volcanism marked by huge outpourings of lava in Siberia. It is not certain that this volcanism triggered the extinction, but the timing certainly looks suspicious and, if the volcanoes truly were the guilty parties, it seems that this event was not influenced by any astronomical phenomenon at all. Once again, the Oort Cloud is declared innocent.

The concept of devastating comet showers from a disrupted Oort Cloud is certainly a widely held scenario, even if not every extinction event can be laid at its feet. Comet showers must occur. This much is clearly indicated by the tenuous nature of the Sun's gravitational hold on the Cloud's most remote members and the fact that encounters with passing stars, groups of stars, and giant molecular clouds cannot be avoided forever. But just how intense are the resulting comet showers? Are they "storms", or merely passing splashes?

The answer depends, in part at least, on the population of the Oort Cloud. The most often quoted figure for the numbers of comets out there is 10^{11} or roughly the same as the number of stars in the Milky Way galaxy. Some astronomers raise the number by a further order of magnitude, or even higher.

The reason for these high figures is the constant stream of comets whose orbits trace back to Oort Cloud distances. Several are discovered every year and when the probability of an object being flung inward from the Cloud actually coming within range of our telescopes, the chance that such an object will in fact be found plus the estimate that about 50% of Oort Cloud comets perturbed out of their Cloud orbits are sent the other way and head into interstellar space are all taken into account, it is clear that we are dealing with very large numbers indeed. Assuming that the observed influx is a steady state stream persisting over much of the life of the Solar System, some estimate of the number within the reservoir of the Oort Cloud can be calculated. This is where the very high figures come from. Clearly, a major disturbance to a population of that size would surely result in a comet storm with potentially devastating effects on Earth's environment.

Yet, there is a niggling problem. The "new" comets that we observe today come in on orbits very close to the escape velocity of the Solar System. The Sun's hold on them is weak and even quite small perturbations by the planets can change their orbits significantly, either pumping them up to escape velocity or pulling them down into long-period ellipses. If a comet ends up in an ellipse, the aphelion point of its new orbit is most likely to fall well short of its original position in the Oort Cloud. Therefore, comets returning on their second trip to the inner Solar System will be distinguished by smaller (though still very large) elliptical orbits and shorter (albeit very long) orbital periods.

We *do* find comets like this, but not nearly as many as might be expected if the present number of "new" ones is typical of the steady state influx over the life of the Solar System or even, for that matter, over the past few millions of years.

Even eliminating the ones that are dynamically pumped up after their approach to the Sun and go flying off into the wider realms of the Galaxy, as well as those that clearly break up and disappear before our eyes, *plus* the few that do manage to return to Oort Cloud distances and are either shifted into other orbits by a subsequent encounter with a passing star, *or* that do manage to return, (albeit in orbits indistinguishable from "new" comets) the discrepancy remains disturbing. So large in fact, that Jan Oort – in whose honor the Cloud is named – was led to postulate an attrition rate as high as 90% for "new" comets. While it is true that some "new" comets (especially small and faint ones that pass within the orbit of Earth) do indeed have a greater tendency than "older" ones to disrupt and vanish, these are not typical of the class as a whole. Unless most "new" comets continue to crumble and break apart after they recede beyond the reach of our telescopes, Oort's estimated attrition rate is way too high.

The discrepancy between the numbers of "new" and "second-time-and-subsequent" comets may also be explained if the assumption that the contemporary number of the former is well above the steady state influx. In other words, we might be living through a comet shower; not a storm, but nevertheless a significant influx. If that is true, the Oort Cloud may be a lot less populous than is widely thought and comet showers might not be as dangerous as we have been led to believe. After all, if we are experiencing one today, we seem to be managing quite well!

Given this argument plus the apparently different causes of the great P-T and K-T extinction events any simple association between extinction events and disturbances of the Oort Cloud must be viewed with a good deal of skepticism. That is not to say that very major disruptions of the Cloud (as, for instance, the effect of a passage through a very dense galactic region) may not trigger unusually intense and dangerous comet showers, nor does it mean that some extinction events might truly have been triggered by the impact of a comet, whether during a shower or randomly. But it does imply that each event should be looked at as an individual occurrence. The history of Earth is complex and the balance of climate delicate. On the geologically short run at least, disastrous changes in climate can and sometimes do occur and it would be unwise to assume that all must have similar causes.

Indeed, some of the less dramatic ones may have their origin in the most familiar astronomical object in the sky. The one on which our very existence on this planet most directly depends!

The Variable Sun

"Variable star" is a term well known to astronomers, both professional and amateur. Monitoring "variables" is, indeed, one of the more popular and useful endeavors of amateur astronomers and can be a fascinating endeavor even for those whose telescopes are not large and who still prefer visual to CCD observing. One class of variable star – Cepheid variables – played a very important role in the measurement of cosmic distances, thanks to their bright intrinsic magnitude plus a remarkable relationship between their intrinsic brightness and the periodicity of their variations. By timing their fluctuations in brightness, their intrinsic brightness, and therefore their distance, can readily be calculated. Simply identify a Cepheid variable in another galaxy, monitor its period of variability and the distance of the host galaxy can be quickly computed.

Not all variable stars are as regular as Cepheids, but all have their secrets for astronomers to unlock; secrets that ultimately tell us more about the nature of stars in general.

Just by calling some stars "variable" we imply that others are not, however this is more a matter of degree than an absolute distinction. Strictly speaking, virtually all stars are "variable" to a certain degree. The ones we call by that name are simply the more extreme examples; mostly stars that are passing through a markedly unstable phase of their evolution.

Our own Sun is one of the least variable of stars, which is a very good thing for us! As already mentioned, it appears to be stable even by the standard of its own very stable class. Yet, it is not *completely* without variability.

The best known cycle of solar variability is the so-called sunspot cycle. Sunspots are disturbances within the Sun's photosphere (literally "light sphere", in a sense the "surface" or "face" of the Sun visible to us) that are somewhat cooler and therefore less bright than their surroundings. Although we give them the somewhat diminutive title of "spot", there is really nothing diminutive about them and large ones could easily swallow several Earths. They may occur singly or in groups of varying sizes and may last for periods of less than 1 day to greater than a week. An especially long-lived one observed in 1943 was tracked for about 6 months (!), from June of that year right through to December.

In the year 1848, Swiss astronomer Johann Rudolph Wolf began making daily measurements of sunspots, taking into account both the number of spots visible at any one time on the face of the Sun and the number of groups into which these spots clustered. These, and subsequent, measurements revealed that the numbers vary over time in a more or less periodic cycle of about 11 years duration. This period is approximate rather than strict and is apt to vary somewhat between cycles.

This (approximately) 11-year solar cycle shows in more than just sunspot numbers. It is a cycle of solar activity in general; in short, it is a cycle of the variability of our star, the Sun. Thanks to the development of space-based monitoring of the Sun's output of energy, we are now aware that the Sun is very slightly "brighter" (in the sense of radiating more energy, not just visual light) at the time of sunspot maximum. The variation is only very small – in the order of just 0.1% of the Sun's average level of output – but it does mean that the Sun is not completely unvarying as the term "solar *constant*" (used to describe the energy output of the Sun) appears to imply. This variability of the "Constant" is something that must be remembered (but appears at times to have been forgotten) in mathematical models predicting the infamous "global warming" trend of recent decades.

It might be supposed that the Sun should be a little *fainter* at times of sunspot maximum. After all, more of the hemisphere facing Earth is covered by less luminous sunspots than it is during times of low sunspot activity. How can a "spotted" Sun be brighter than a "clear" one?

The resolution of this apparent paradox lies with the margins surrounding the darker sunspots. These regions are actually hotter and brighter than average regions of the photosphere and the combined effect of these not merely cancels out, but actually exceeds, any dimming effect due to the presence of the spots themselves.

Sunspots were known long before Wolf began his daily counts. Indeed, large naked-eye spots were recorded by the ancient Chinese astronomers and the ups and downs in sunspot numbers can at least be approximately traced back into the 1600s. But once the cyclical nature of sunspot frequency became widely known and accepted and an approximate period determined for it, all sorts of supposed correlations were proposed between the solar cycle and events on Earth. The weather has often been suspected of showing a solar-cycle influence. This is not an unreasonable notion, and will be taken up shortly. But other alleged correlations are far more dubious. Sunspots are sometimes blamed for influencing us psychologically, causing us to be more aggressive in years of solar maximum and more cautious and conservative during times of minimum. According to proponents of this view, there is a greater tendency for wars to break out during years of high solar activity as international leaders become more "reckless." Domestically, if governmental elections are held during the years near solar maximum, left wing political parties have a greater tendency of being elected, while conservative and generally right wing groups are likely to be voted to power during times of quit solar activity. As far as I am aware, no politician of either wing has tried manipulating times of elections to take advantage of this alleged tendency. (In the writer's home state, a recent election held as the Sun was starting to activate again after a very low minimum, saw a mildly left-of-centre government of 16 years swept away by a veritable conservative tsunami. No political commentator blamed sunspots for the debacle). Then there was the suggestion that sunspots were behind the economic trade cycle. True, the fluctuations of economic boom and bust follow a periodic trend roughly commensurate in duration with the solar cycle, but any association surely ends there. Economic theorists find far more plausible causes for the trade cycle, a lot closer to home than the Sun.

Nevertheless, despite deservedly skeptical reactions to the more bizarre "correlations," the suggestion that the solar cycle can influence Earth is not an unreasonable one. And if it is capable of influencing Earth, the most likely place to detect that influence is in weather, climate, and atmospheric phenomena generally.

One correlation that is readily apparent is the connection between solar activity and the spectacular high-latitude luminous phenomenon known as the aurora. We will, however, leave further discussion of this until the following chapter.

Another correlation, albeit one which leads into controversial territory, involves the solar cycle and weather. Here again, some proposals have been rather extreme and just about every type of poor weather outbreak has been laid at the feet of sunspots at one time or another. Yet, the baby must not be thrown out with the bathwater. A comparison of global surface air temperatures and sunspot numbers between 1960 until 2010 does indeed reveal a clear correlation. Years of sunspot maximum, other things being equal, were warmer than years of sunspot minimum. Of course, other things are not always equal. A strong El Nino effect or a major volcanic eruption shedding a veil of dust through the stratosphere can seriously throw the correlation out of sync, but in the absence of such events, the pattern returns.

Now, looking more deeply into the Sun's variability, we might well ask if the 11-year cycle is simply part of a longer one.

The answer appears to be "Yes." From sunspot data gleaned from records dating as far back as 1610, Judith Lean and colleagues from the *Naval Research Laboratory* plotted solar irradiance from that year until 2000, uncovering some very interesting results.

First of all, taking a broad brush approach, solar irradiance has risen from the early 1600 to the present day. Indeed, according to these results, Earth has been receiving more heat and light from the Sun since around 1940 than at any time during the last 330 years – and it has continued to rise ever since, albeit slowing somewhat since circa 1990. The 11 year cycle is superimposed on these longer trends, except for one period about which more will be said in a moment.

The trend has, however, not been a steady one. Very far from it. In fact, solar irradiance fell steeply from the early 1600s until the middle of the seventeenth century. It then bottomed out in the early 1700s during a period known as the *Maunder Minimum*, normally dated from 1645 until 1715. This is the period referred to briefly above when the 11-year solar cycle disappeared altogether. The "Minimum" was followed by a steep increase in solar activity until about 1770, before plunging again into the *Dalton Minimum*, lasting from 1795 through until 1825. The shorter cycles did not disappear during this second (and less severe) minimum, but they became distorted, with long declines on the downward slope of the Dalton Minimum and greatly lengthened waxing and shortened declining phases on the recovering slope. The maxima of the shorter cycles during that time were very low and brief. A small maximum of the average solar irradiance was reached around 1840, albeit lower than the one immediately preceding the Dalton (which peaked in 1780), followed by a very shallow decline until the early 1900s. This in its turn gave way to an upturn from about1910. The upward trend continues to the present day, lifting the solar irradiance to levels well beyond anything seen previously during the period under consideration.

Now, it is not unreasonable to think that these variations in solar output – slight though they be in the overall scheme of things – would be reflected in Earth's climate and, indeed, we do find evidence to this effect. The most striking correlation between what the Sun is doing and conditions here on Earth involves the period of low solar activity known as the Maunder Minimum. As we remarked, solar activity effectively bottomed out for a period of some seven decades and few people think it coincidental that this time was one of quite severe global cooling, especially in Europe. Although these years certainly fell far short of true ice age conditions, the period has been called the *Little Ice Age* as a mark of just how noteworthy the cooling became. Rivers that had not frozen over in the corporate memory of communities close to them suddenly became ribbons of ice in winter time. The Thames in London, for instance, had people skating on it; certainly not a normal winter's pastime in that part of the world.

Yet, it is not immediately clear how a slight diminution in the Sun's output could cause such severity of freezing. This has led some to downplay the role of solar cycles in climate changes, but the Little Ice Age surely stands as a good indication that some association exists, even if the full story is not easily teased out.

One interesting suggestion as to how solar activity impinges on terrestrial conditions involves a somewhat indirect chain of events involving cosmic rays. These so-called "rays" are really sub-atomic particles which enter the Solar System from interstellar and even intergalactic space. They are generated by very energetic events elsewhere in the universe and in many instances have been accelerated to highly energetic levels by galactic magnetic fields. About 90% consist of protons (nuclei of hydrogen atoms) with a further 9% consisting of alpha particles (one proton and one neutron – the nuclei of helium atoms). Electrons make up 1%.

As these particles enter the Solar System, they run into the stream of protons boiling outward from the Sun, known as the "solar wind." This "wind" is, of course, "blowing" in the opposite direction to that of the incoming cosmic rays, and carrying along with it a magnetic field against which incoming charged particles must battle. During the years of greater solar activity near the crest of the solar cycle, this "wind" blows strongest, the magnetic field is strongest, and cosmic ray particles – especially those of the less energetic variety – have a much harder battle to penetrate the realm of the inner planets. As a consequence of this, the cosmic ray flux reaching Earth and the other inner planets varies inversely with the level of solar activity. Cosmic rays are fewer at solar maximum than at solar minimum. Presumably, the crests and troughs of the longer periods of solar activity are also reflected in Earth's level of cosmic ray exposure. It would seem to follow that a period of prolonged very low solar activity, such as the Maunder Minimum, should be one in which Earth is exposed to an unusually high flux of cosmic rays entering its atmosphere. Moreover, the length of time that this influx occurs would far exceed the normal bursts typical of "normal" 11-year solar minima. But could this influx of particles from the depths of outer space really affect our climate?

According to the research of Henrik Svensmark and Eigil Friis-Christensen, the answer to that question may well be positive. In short, just as sub-atomic particles passing through a Wilson Cloud Chamber leave a trail of tiny cloud droplets in their wake, so cosmic rays passing through Earth's atmosphere leave a similar mark. According to these authors, times of low solar activity and high cosmic ray exposure are times of greater cloudiness. More clouds (comprised of small droplets) increase the planet's overall albedo and this in turn cools the lower atmosphere. This may not be the only cooling influence, but it may be enough to trigger "little ice ages" during periods such as the Maunder Minimum.

Unfortunately, although Chinese records of naked-eye sunspots go back many centuries, these observations are sporadic and anything approaching regular data does not extend earlier than the period we have been considering here (i.e. not before the early 1600s). Regular day-by-day records, as we saw, were only taken from the middle of the nineteenth century. This makes it difficult to determine the regularity, or even the existence, of longer solar cycles, but it is reasonable to think that the period since 1600 managed to capture the last section of a decline in activity that terminated in the prolonged trough of the Maunder Minimum and has been climbing toward a new maximum ever since. The years since 2000 might temp us to think that we are now pretty close to the crest of this longer cycle as the increase in solar irradiance, especially pronounced during much of the twentieth century, now seems to be reaching a plateau. We are now nearly 300 years from the end of the Maunder Minimum. If we are near the maximum phase of a cycle, presumably the previous maximum occurred around 300 years prior to the beginning of the Maunder Minimum or around the late 1300s. Is there any reason for thinking that this might be true?

Possibly. If we make the assumption that (for whatever reason or reasons) "grand minima" such as the Maunder Minimum coincide with cold periods and "grand maxima" with warm, it is interesting to note that the period from around 1000 to the 1200s was a time of global warming. This is admittedly a little early for a warm late 1300s, but these estimates are extremely rough and depend on such assumptions as the period of decline to minimum equaling the rise to maximum, the length of the cycle being regular (and we already know the shorter so-called sunspot cycle is not) and even that we have reached or nearly reached the maximum. The present plateau of solar irradiance may not be maximum after all – it may just be a pause – or the maximum may last for 100 years or more. We simply don't know. But we *do* know from tree ring analysis and from historical evidence that these mediaeval centuries were warm. Those who today point to the warming of Greenland as evidence of climate change doom seldom remind us that about 1,000 years ago Greenland was literally a *green land*. Greenland farmers raised crops and brewed beer there around 1000 AD, so the island has quite a deal of warming to go through yet if it is to return to those long ago days of summer.

Climate Change Today: Man-Made or Natural?

This brings us to ask something that is controversial, heretical, politically incorrect and in general inviting of derision, abuse and even the odd four-letter word. Namely, "Does contemporary climate change arise, not from human activity, but in whole or in part, from a rising solar irradiance or some other "natural" cause?"

The question looks harmless enough, but in the writer's own country and, I suspect, in most other places as well, just raising the possibility invites such vitriol as to be worthy of a sociological study in its own right. This reaction is not so much evident in private discussions (where, perhaps significantly, many are found to agree with the skepticism behind the question) but in the public media. Newspaper columnists are fond of contrasting "skeptics" and "scientists" (apparently oblivious to the fact that many of the first category also belong to the second!) and then painting the "skeptics" as ranting rednecks who stubbornly deny the obvious. Typically, they avoid mention of people such as David Bellamy, an environmental activist whose conservationist credentials are as green as an Irish meadow, but who has moved from being a global warming believer to a thorough skeptic who was even heard to utter the highly heretical statement "carbon dioxide is not a pollutant"! Someone who has been arrested for campaigning for the environment, as Bellamy has been, is difficult to dismiss as an ultra-conservative buffoon. No doubt, that is why his ideas on the subject are quietly passed over.

The thesis that our burning of fossil fuels is the sole cause of climate change and the assumption that this can only end in doom has been used in a barefaced manner by firms trying to frighten customers into buying their products, or at the very least to put potential consumers on a such a guilt trip that any alternative should appear an immoral choice. Lest the reader think that this is too harsh a judgment, I can only rest my case by calling to mind a television commercial that begins "OUR PLANET IS DYING!" and then proceeds to inform viewers that they can do their bit to stave off our planet's demise by purchasing their product. Another commercial has a child answering someone who says "You can't hurt the world" with "You can if you don't buy [the advertised product]." And there is always the slogan "buy [our product] and you will be helping the planet." And we wonder why there is so much depression amongst youth and why so many young folk have a sense of hopelessness about the future!

For the record, the writer is not a member of the political far Right (actually, I am somewhat to the Left of Centre), strongly support the control of pollution and am not opposed to taxation schemes that make large scale polluters pay for improving the environment. However, the issue of global warming, climate change or call-it-what-you-will is a scientific, not a political, one and it is a great pity that the issue has become so emotionally changed, so politicized, so factionalized and so prone to generate more heat than light in any discussion of the topic. An ancient Greek philosopher once said that truth lies at the bottom of a deep well. Alas, the deep well at whose bottom the truth about climate change lies has been so muddied that it has become all but inaccessible.

So, is there any hope of diving into that deep and turgid well and recovering the truth about climate change? Maybe, but if the conclusion does not agree with the prevailing politically correct interpretation, it is doubtful if many will listen.

Still, let's just look at a few facts and reports that threaten to emerge as inconvenient truths for proponents of anthropogenic (human-caused) climate change.

First, global warming appears to have taken place on Mars in recent decades. This is evident from orbiter images of steadily shrinking knolls and low mesas of ice, especially at high southern latitudes. Also, ground depressions believed to have been caused by melting permafrost have enlarged since the first orbiting spacecraft imaged wide regions of the Martian surface. Now, just because Mars and Earth are neighboring planets, this does not necessarily mean that the engine driving warming on both is the same. Clearly, anthropogenic factors cannot be responsible for any global effect on Mars, but it is also difficult – impossible more likely! – to see how cosmic-ray induced cloudiness could be responsible either. Clouds are not something that Mars produces in abundance at any time! A common cause might be the direct influence of increased solar irradiance, but it is hard to understand how such small variations could bring about noticeable temperature changes at the surfaces of planets. Nevertheless, the example of Mars demonstrates that Earth is not alone in experiencing climate change, and at the very least cautions us not to always assume a large human contribution to global warming.

Secondly, is the degree of global warming that has actually taken place equivalent to that expected from the continued rise in the carbon dioxide content of Earth's atmosphere?

The popular presentation of anthropogenic climate change implies that all global warming since at least the beginning of the twentieth century has been caused by carbon dioxide emission from human industry. But that is not quite what the scientists themselves are saying. Global warming is not caused by anthropogenic factors alone and the warming trend prior to 1970 can largely be accounted for by other means. It is the period since 1970 where, they argue, human based factors overwhelm the purely "natural" ones and which gives rise to such ominous indications of what is to come. But how much warming has really taken place since 1970, and is it accelerating as many alarmist voices keep telling us?

The problem is, fluctuations in temperature occur from year to year and it is not always an easy thing to draw an accurate average curve through a whole range of yearly peaks and troughs. According to a study of global surface temperatures from the *Hadley Centre for Climate Prediction and Research* and the *Climate Research Unit of the University of East Anglia* in the United Kingdom, a warming trend of 0.45° took place from 1962 until 2003, but this was anything but a smooth upward slope. In fact, between 1962 and 1965, the global temperature *fell* by 0.2°, then rose 0.1° until 1974, before tumbling again to the 1965 level. (As an aside, the writer recalls that in 1965 heavy snow fell within sight of his home; something that had not happened for 70 years. Then, in 1975, the same thing happened again.)

From 1974, the average trend has been upward, albeit with significant undulations; 1982 averaging 0.1° warmer than the 1962 starting point, but 1985 going back to base again. The year 1998 was a very warm peak, about 0.7° up from 1962, but that was also a strong El Nino year and it is believed that this event was largely responsible for the year's high temperatures. By contrast, 2009 dropped down again to the 1982 level. Taking the average curve from 1998 through to 2010, global temperatures reached a plateau from 2003 and actually fell by about 0.1° between 2007 and 2010. It is this apparent steadying, even slight falling, of the average global temperature since the turn of the century, even though carbon dioxide continues to increase, that has added to skepticism about the "official" position on climate change.

Climatologist Professor Richard Keen is concerned that the small trends which supporters of anthropogenic climate change model try to tease out of the data is simply attempting to detect too small a signal amongst too much noise. The data is indeed very "noisy." There is a lot of scatter of data points as global temperature fluctuates (from many causes) from year to year and month to month. Data points at different parts of these sharp short-term fluctuations could be connected to "prove" just about anything one wished.

Keen wonders if so weak a signal as that used to support a serious rise in global warming during the last four decades is real. The fact that supporters of the anthropogenic model and their critics can get different signals from the same data (each supporting one's own position) adds weight to his concerns.

Keen draws attention to a factor that is often overlooked in public presentations of the climate change issue, namely, atmospheric transparency. Major volcanic eruptions are known to eject great quantities of fine dust into the stratosphere causing a level of opacity greatly exceeding that of what we might call "normal" periods or periods free from major volcanic activity. A telltale indicator of the presence of such a stratospheric dusty layer is the lightness or darkness of any lunar eclipse occurring during these times. An eclipse occurring during times of clear stratosphere will be "light." There will still be a great diminution in the light of the Moon of course, as the only sunlight reaching its surface during the total phase of any eclipse will be that refracted through Earth's stratosphere. This is necessarily much fainter than direct illumination by the Sun, even when the stratosphere is as clear as it can possibly be. But a hazy, dusty, stratosphere makes a marked difference. For instance, during the eclipse of December 30, 1982, occurring while the stratosphere was heavy with dust from the El Chichon eruption, the Moon was some 400 times fainter than it had been during a "clear stratosphere" eclipse of 10 years earlier. An even greater difference was evident between the 1972 and 1963 eclipses. The total eclipse of December 30 of the last-mentioned year saw the Moon several *thousand* times fainter than its dimmest point at an average eclipse. Normally, the eclipsed Moon is conspicuous with the naked eye even throughout the deepest point of totality. Yet, during the 1963 eclipse, this writer lost sight of it altogether! It is not surprising that this was the only time that I have had that experience, as the eclipse of 1963 remains the darkest since the one way back in 1816! The reason why the 1963 eclipse was so dark lies with the eruption of Mt. Agung on the island of Bali earlier that same year.

Project 3: How Bright Is the Eclipsed Moon?
(1 – In terms of Stellar Magnitude)

(Readers of my Weird Astronomy will already know this – and the following – project and to them I apologize for the repetition. Still, as I cannot assume that all readers of the present book will be acquainted with the former one, these projects are repeated here for these new readers).

Because of the large angular size of the Moon, it is not easy to compare its brightness during eclipse with that of a star. Moreover, as it will be brighter than most of the stars (except for the darkest eclipses), there will be few with which it can be compared. These difficulties can be largely overcome if the eclipsed Moon is viewed through the reverse end of a pair of binoculars and this reduced image compared with that of a star seen by naked eye.

However, a little simple math must be done in the process.

Looking through the "wrong end" (the object lens) of a pair of binoculars reduces the diameter of the Moon's image by the same factor as it would be magnified if we looked through the "correct" end. In other words, if the binoculars are ten magnification (e.g. 10 × 50), looking through the reverse makes it appear ten times *smaller* than by naked eye. Similarly, the same procedure using a 6 × 35 or 4 × 30 binocular will reduce the apparent diameter by six and four times respectively. Likewise, the area of the image will be reduced by the square of these values; i.e. 100 times for a tenfold reduction in diameter, 36 times for a sixfold reduction of the diameter and 16 times for a fourfold.

(continued)

Project 3: (continued)

Now, the brightness of the reduced image is a function of its area. Therefore, if the area is reduced by a factor of 100 (as through the reverse end of a pair of 10×50 binoculars), the brightness is also reduced by that amount. Similarly, if the area is reduced by 36 or by 16, the total brightness of the image is reduced from the naked-eye value by the same amount.

The brightness of astronomical objects is given in terms of a magnitude scale in which each step of one magnitude represents a difference in brightness of 2.512 times; the larger positive values being fainter. Thus, a star of first magnitude or magnitude +1 is 2.514 times brighter than a star of second and 2.514 times fainter than one of zero magnitude. A star of magnitude –1 is 6.46 (2.541^2) times brighter than one of magnitude +1 and a star of magnitude 5 is $2.541^5 = 105.9$ times fainter than one of magnitude zero.

Therefore, suppose the reduced image of the eclipsed Moon as seen through a reversed pair of 10×50 binoculars compares closely to the naked eye image of a second magnitude star. Because the reduced image is 100 times fainter than the true naked-eye image, in terms of the magnitude scale it will be just a shade under five magnitudes fainter than its true naked-eye image. For this purpose, rounding off the magnitude difference to 5 is quite accurate enough. Therefore, if the "reduced" image is estimated as +2 (i.e. equal to a star of that brightness), the "true brightness" will be $2 - 5 = -3$. The true brightness of the eclipsed Moon is therefore –3; a rather bright eclipse.

If, on the other hand, the eclipse was observed through a pair of reversed 4×30 binoculars the magnitude difference would be approximately 3 ($2.514^3 = 15.9$). If the reduced image was estimated as magnitude 4, the true brightness would be just +1; a truly dark eclipse.

Since the eruption of Pinatubo in 1991, there was no major ejection of fine volcanic dust into the stratosphere until 2011 and all lunar eclipses between a relatively dark one in early December 1991 and mid June 2011 were bright, indicative of a clear stratosphere. The June 15, 2011 eclipse was quite dark thanks to a heavy ejection of volcanic dust earlier in the year and continuing right through the time of the eclipse itself.

Now, it is well known that the presence of large amounts of stratospheric dust cools the Earth's lower atmosphere. The downturn in global temperature following the Agung blast in the early 1960s has already been noted, and similar ones are evident in 1991 and, to a lesser degree, in 1982. And before any of these relatively recent occurrences, 1816 and its remarkably dark eclipse was followed by "the year without a summer"; a serious cold spell throughout Europe. Conversely, it must follow that years of clear stratosphere will be warmer (other things, as always, being equal). Keen calculates that the "clear" period from 1998 should yield mean global temperatures about 0.2° warmer than the global averages across the previous four decades.

Project 4: How Bright Is the Eclipsed Moon
(2 – According to the Danjon Scale)

The brightness of a lunar eclipse may also be estimated according to a five-point scale proposed by French astronomer A. Danjon. On the Danjon Scale, the brightness of an eclipse is measured, not in terms of stellar magnitude, but according to the color of the eclipsed Moon and the clarity of detail on its disk. Readings on the Scale are given in L values as follows;

L = 0 Very dark eclipse with the Moon almost invisible, especially at mid eclipse.
L = 1 Dark eclipse with a grey or brownish coloration. Details on Moon distinguishable only with difficulty.
L = 2 Deep red or rust-colored eclipse. Central shadow very dark, while outer edge of umbra relatively bright.
L = 3 Brick-red eclipse. Umbral region usually shows a bright or yellow rim.
L = 4 Very bright copper-red or orange eclipse. Umbra has a bluish and very bright rim.

Estimates in between these values (e.g. L = 3.5) can be made when appropriate, as the scale – like most scales of this type – covers a gradation rather than strict categories.

Although volcanic dust is the major contributor to global atmospheric clarity, it is not the only one. Ironically, since environmental awareness led to a cleaning up of particulate emissions from factories in European countries, the rise in atmospheric clarity from a cleaner environment has been a minor contributor to increased temperatures over that part of the world. The irony of green policies adding to global warming might be amusing, but nobody suggests that we go back to the days of smoke, smell and killer smog. Any calculated slight rise in temperature is far less concerning than breathing clouds of burnt coal dust!

In a presentation in 2003, Keen estimated that about one half of the rise in global temperature since 1980 was due to the clearing of the upper atmosphere of volcanic dust. That left just 0.1% to be accounted for by warming from other causes. These, according to Keen, include a variety of factors such as solar variability, El Nino, oceanic variability, albedo changes … and maybe a little anthropogenic greenhouse warming. The latter is not, in Keen's opinion, nearly as significant nor as menacing as popular thought makes it out to be.

If Keen is correct, the warming due to increased solar irradiance has been slightly enhanced during the past couple of decades by a clearer stratosphere and other minor contributions; including, perhaps, a little anthropogenic carbon dioxide greenhouse warming. But, as Mark Twain said about reports of his death, the importance of the latter has "been greatly exaggerated" in most presentations of the subject.

Even so, the warming trend (no matter what its cause) has not stopped some recent years from being colder than those earlier in the decade or even in past decades. The year 2007 was colder (on a global average) than any other year in a decade and there were unusually cold winters in both northern America and Europe in 2010. Trends can be very difficult to follow through such large year-to-year variations.

Popular press coverage of global warming and climate change trends of recent years has also drawn attention to events such as severe weather outbreaks, glacial melt and the like, as evidence that a dangerous degree of warming is taking place. The implications are that as the warming trend continues, severe events of

this nature will increase and may even endanger the fabric of society as we know it. Statements such as "the hottest day on record" or "the most severe hurricane on record" clearly sound ominous, but we have to admit that there is a tendency to interpret "on record" as being synonymous with "ever"; something which is certainly not the case. Meteorological records do not go back very far on the scales of human history. It is not unfair to say that the records we have at our disposal only cover (part of) the period during which Earth's climate pulled out of the cold spell of the 1600s. Tree rings and so forth extend this back further into the past; in a general sort of way. But we cannot truly say just how "typical" is our period of the planet's climatic history as a whole. Have we been fooled by living in a slightly more benign epoch than other phases of history? This question has been asked more than once, but not yet satisfactorily answered.

Moreover, is it strictly true to say that violent weather events have increased in recent years; since, say, 1970 when the "official" position says that anthropogenic global warming overtook "natural" climate change trends?

Once again, Richard Keen is skeptical. As part of his argument, he lists the following inconvenient facts; "inconvenient", that is, for those who insist that violent and unusual weather is a recent phenomenon caused by global warming:

Alaskan glaciers melt at the rate of 1 mile per year! Clearly an ominous sign of global warming ... except that this occurred in 1900!

Alaska experiences a heat wave with temperatures soaring above 100°F north of the Arctic Circle ... in 1915.

Another heat wave, this time in North Dakota, sends temperatures to 120°F ... in 1936.

Boston experiences a very rare (for that part of the world) category 5 hurricane ... in 1938.

Los Angeles is hit by a hurricane (very rare in that city) resulting in massive destruction ... in 1839.

That same city is struck by a series of tornadoes (also very rare in California) ... in 1983.

The USA experiences its warmest year on record, with 20 states setting all time heat records ... in 1934.

This does not exhaust Keen's list, nor is his list any more than a sample of what might justifiably be called extreme weather events. We might add, for instance:

The hottest day yet officially recorded anywhere on the planet occurred on September 13, 1922 when the mercury soared to 136°F at El Azizia in Libya.

The highest temperature recorded in Israel – 129°F on June 21, 1942.

The highest European temperature – 122°F at Serville, Spain on August 4, 1881.

The record hottest for North America – 134°F at Death Valley in 1913 and for South America, 120°F at Rivadavia, Argentina on December 11, 1905.

Although they do not *prove* anything, these few entries are enough to call into question much of the sensationalist talk that has been flooding the airwaves in recent years.

Such questioning invites a healthy skepticism, not a closed minded rejection of everything stated by those of opposite persuasions. It is here that the political associations muddy the water and make things hard for the inquirer to get a fair picture of the situation. Name-calling has no place in science but unfortunately it is not uncommon in the climate change debate (as remarked earlier) and is apparent to anybody with even the most limited exposure to "discussions" of the subject. Unfortunately, both sides often have agendas other than providing a fair and balanced overview of the issue.

Whatever the situation regarding climate change, we human beings must be ever aware that we have a responsibility to keep this home planet of ours a clean and attractive place fit for the habitation of future generations. Climate change skepticism must never be used as an excuse for environmental neglect. If government taxes on carbon help to modify air pollution, then they should be supported, even by those who might not share the opinion that they are "saving the planet" from a global warming catastrophe. Surely, we do not need to be terrified by an image of Earth as Hell to encourage us to try our best to make our home world just a little more like Paradise!

3. Glowing Skies

Dancing Lights in the Polar Night

Although rarely seen at low latitudes, the skies closer to the north and south magnetic poles of our planet are frequently lit by one of the most spectacular of nature's light shows; the aurora or polar lights.

Auroras assume a wide range of forms, from dim greenish glows to spectacular curtains of light and displays of beams and rays that put our laser shows to shame.

Fig. 3.1. Aurora borealis from space, as imaged by the International Space Station 6 team (Credit: NASA)

D.A.J. Seargent, *Weird Weather: Tales of Astronomical and Atmospheric Anomalies*, Astronomers' Universe, DOI 10.1007/978-1-4614-3070-4_3, © Springer Science+Business Media New York 2012

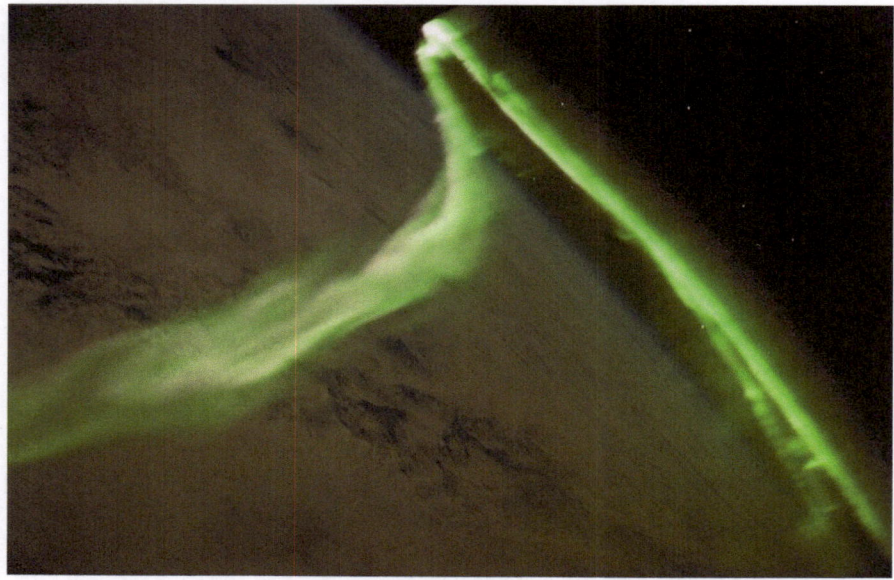

Fig. 3.2. Aurora during a geomagnetic storm probably caused by the solar coronal mass ejection of 24 May, 2010. Photographed from the ISS (Credit: NASA)

Many displays, especially as seen at lower latitudes, are relatively mild affairs; a glow over the northern or southern horizon (depending upon which hemisphere you are in), perhaps suggesting a large fire just over the hill or even a glow from the lights of a distant city. The glow rises and falls, as if a fire blazes more fiercely upon reaching fresh fuel and then dies down again as this is consumed. Sometimes we might see what look like racing clouds illumined by a greenish glow from some light source at ground level. Except that the "clouds" have dark undersides and luminous "tops" and their rate and direction of travel may pay no heed to either the velocity or direction of local wind.

At other times however, the glow is anything but quiescent, and can be mistaken for nothing other than its true character. A huge arc of brilliant luminescence spans the sky. Or something like a billowing and rippling drapery of multi-colored light sways and waves through the heavens. Often, what begins as an arc bursts into a frenzy of activity, dividing into short rays in one place and pulsating patches of diffuse light in another. Sometimes long shafts of light, looking like powerful searchlight beams or laser lights

shoot upward from the horizon. For observers fortunate enough to be near one of Earth's two magnetic poles during a vigorous display, light beams having a multiplicity of colors may shoot up from points all around the horizon and converge overhead in a display that one observer of an especially spectacular example likened to being inside a luminous circus tent. This convergence of rays overhead is known as a *corona* and marks the climax of the display. The convergence overhead is an effect of perspective, like the apparent convergence with distance of parallel railway tracks, but that does not make it any less spectacular. Sometimes waves of light, surging upward from the horizon, appear to fan weaker luminous areas into greater activity, like a wind blowing across the embers of a dying fire.

Then, as quickly as the display sprang into action, it collapses. What had been an incredible light show only minutes earlier, fades to nothing more than a pale glow on the distant horizon.

Fig. 3.3. Red & green aurora over Fairbanks, Alaska, from an old print (Credit: Mila Zinkova)

Over the centuries, many ideas have been put forward to explain these displays. To the traditional Inuit people, they were the spirits of unborn children playing in Heaven as they waited to be born into the world. The Cree referred to them as the "Dance of the Spirits" while in Mediaeval Europe, they were seen as a sign from God, rather as the Europeans of the Middle Ages understood comets. Indeed, in chronicles of the time, aurora and comets are not always easy to distinguish as, for instance, Ambroise Pare's oft-quoted and lurid description of a "comet" in 1528, that is nowhere else recorded and was almost certainly a bright auroral display. It is worthwhile quoting his description, even though many readers will already know it. But in deference to the true nature of what he is describing, I will here take the liberty of replacing Pare's use of the word "comet" with "aurora." The description, though still luridly fantastic, will make better sense. With this single alteration, Pare writes:

> So horrible was it, so terrible, so great a fright did it engender in the populace, that some died of fear, others fell sick ... This [aurora] was the color of blood; at the summit of it was seen the shape of a bent arm holding a great sword as if about to strike. At the end of the blade were three stars. On both sides of the rays of this [aurora] were seen a great number of axes, knives, bloody swords, among which were a great number of hideous human faces, with beards and bristling hair.

Clearly, a lot has been read into this, just as we see faces in clouds and psychologists invite us to see all sorts of strange things in ink blots. But it is also not too difficult to see in most of the "axes, knives and bloody swords" regular auroral features, even if we can't be too dogmatic about which specific feature is which. Nevertheless, the bent arm sounds as though it may have been a portion of an arc, the sword an auroral beam, the knives short rays and the axes and faces luminous patches viewed with an imaginative eye.

At later times, when natural explanations were sought for these displays, the belief that they were caused by sunlight reflecting off the polar ice cap became popular. This had some superficial appeal, in so far as most of the world's population needed to look north – in the direction of the North Pole – to see the lights. Also, as the southern realms started to be opened up to European explorers, the existence of the southern counterpart became better

Fig. 3.4. Aurora over Bear Lake, 18 January 2005 (Credit: U.S. Air Force)

known and this too was seen in the direction of the (south) pole. But did the polar lights ever appear in the dead of winter, when the pole for the winter hemisphere is in darkness? Yes, they did!

Also, although only recognized later, the "polar" lights are not strictly speaking "polar" at all. We now know that they mostly occur in rings that lie just inside the Arctic and Antarctic circles, around 65–72° latitude. From the far northern islands of Spitsbergen (around 80° north) one typically needs to look south to see the northern lights! Auroras closer to the pole do happen occasionally, but they are normally faint; even below the threshold of the naked eye.

Also involving snow and ice, but avoiding the pitfalls of the reflection hypothesis, was a rather ingenious idea put forward by Benjamin Franklin, namely that the northern lights resulted from a concentration of electrical charges in the polar regions intensified by snow and other moisture. Franklin was not too far from the truth in thinking that electrical phenomena were involved, though not in the way he suspected. He also had it right about the electrical potential in snow. As we shall see in the next chapter, odd luminous phenomena and peculiar lightning-like discharges have been reported during snowy conditions. Alas for Ben though, these have no bearing on the aurora!

Fig. 3.5. Aurora australis from Bluff, New Zealand, 1994 (Credit: Paul Moss)

The true cause of the aurora could not have been discovered within the framework of pre-twentieth century physics. While ever the atom was conceived as a genuine "atom" – i.e. as something hard and indivisible – the cause of the polar lights could only remain a mystery.

Once it was found that so-called "atoms" are only the smallest particles of elements and not the smallest particles of matter per se and once it became known that at least some of the subatomic particles could be stripped away from their parent atoms, the stage was set for the discovery of the true mechanism of the aurora.

Auroras are now known to have their origin far away in the outer atmosphere or *corona* of the Sun. (The use of the word "corona" to describe both the climactic phase of an aurora and the outer atmosphere of the Sun is quite coincidental, by the way!). Blowing outward from the Sun's corona, and really forming an extension of it, is a stream of protons or ionized hydrogen atoms – that is to say, hydrogen atoms that have had their electron shells stripped away – and free electrons. These denuded atoms and their

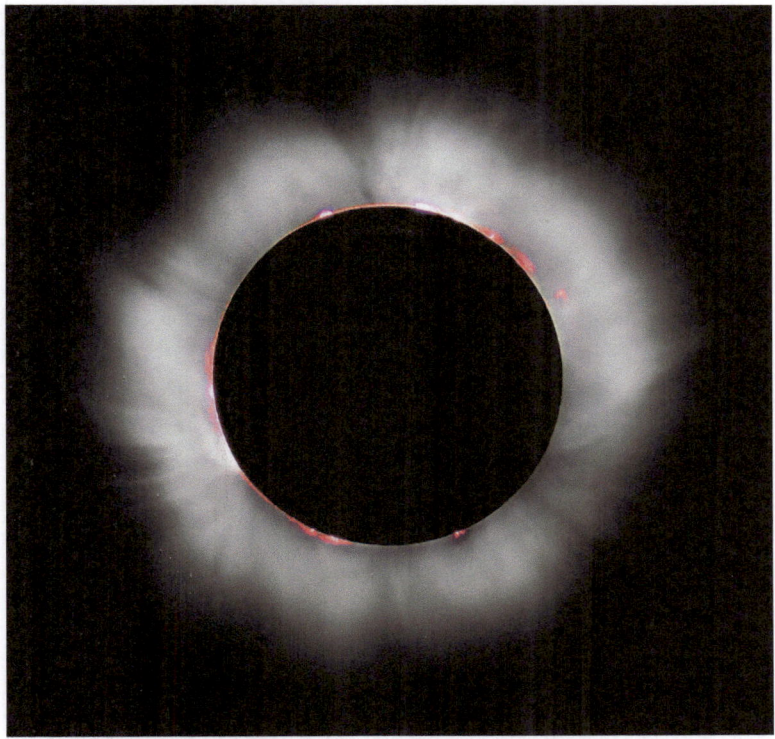

Fig. 3.6. The eclipsed Sun revealing the corona, 11 August 1999 (Credit: Luc Viatour)

stripped-away electrons testify to the extreme heat of the corona. Temperatures there exceed one million degrees, although the material composing it is so tenuous that these temperatures have little meaning in terms of (what we might call) "felt heat." These outflows away from the corona are known as "solar corpuscular streams" or, in more homely terminology, the "solar wind." At Earth's distance from the Sun, this "wind" blows at a speed of 900,000 miles (or 1,440,000 km) per hour. It is a good thing for us that it *is* very tenuous and that our planet possesses a magnetic field capable of diverting much of the wind around the planet itself.

The "wind" is not a neat steady stream. At times the Sun erupts in a display of cosmic fireworks. Flares and ejections of clouds of plasma from the corona create "squalls" within the solar wind that greatly, albeit briefly, enhance its impact upon the interplanetary environment.

Fig. 3.7. Turbulent Sun on 1 August 2010. This extreme UV image from Solar Dynamics Observatory shows the Sun's northern hemisphere during mid-eruption. The event triggered geomagnetic storms and aurora as far south as Wisconsin and Iowa on August 3 (Credit: NASA/SDO/AIA)

The impact of the solar wind upon the magnetic field of our planet creates the Earth's magnetosphere, which diverts the wind at a distance of about 44,000 miles (70,000 km), forming a bow shock some 7,500–9,500 miles (approximately 12,000–15,000 km) further upstream. At the position of Earth, the magnetosphere is around 119,000 miles (190,000 km) wide. Beyond the side of the Earth facing away from the Sun (i.e. the night side of our planet) a magnetotail streams far into interplanetary space like an invisible comet tail.

Were it not for the shielding effects of this magnetic field, the solar wind would erode our atmosphere over time. But as well as shielding us from the solar wind and its adverse effects, the terrestrial magnetic field also traps it in the Van Allen radiation belt. Because the terrestrial field, like all magnetic fields, has two poles

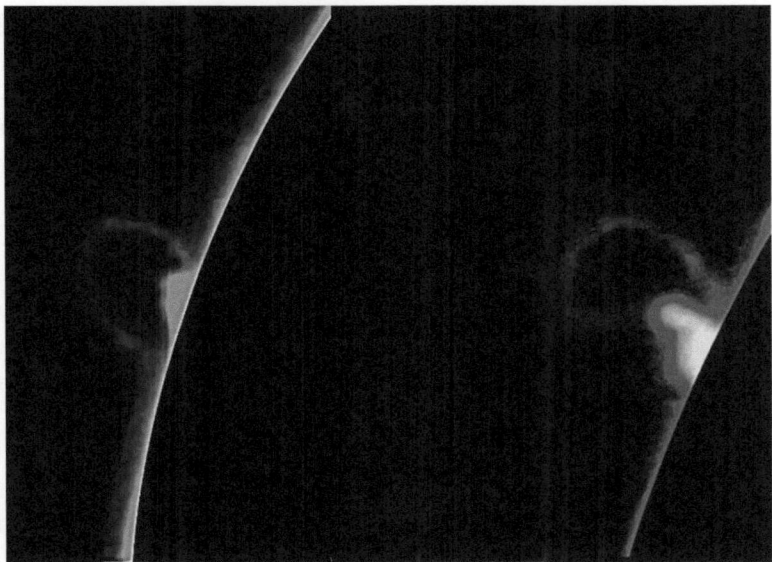

Fig. 3.8. Solar flare and after-flare prominence, 6 August 2007 (Credit: I, Mbzl. Mila Zinkova)

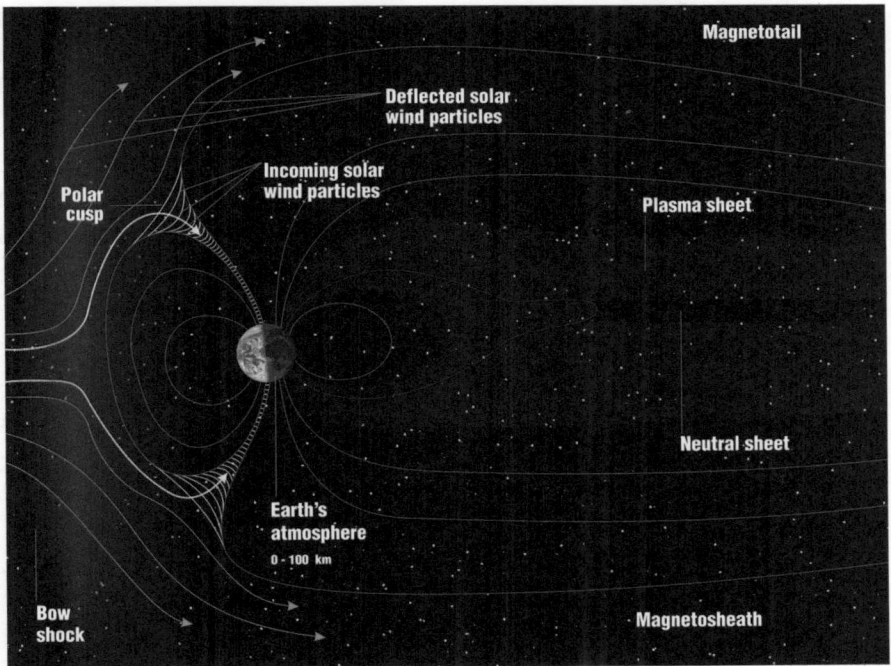

Fig. 3.9. Schematic diagram of Earth's magnetosphere (Credit: NASA, Aaron Kaase)

where the lines of magnetic force concentrate, many of the trapped particles funnel downward toward these magnetic poles and into the upper atmosphere. It is here, at altitudes of between about 50 and 600 miles (80 to about 380 km) that they encounter atoms of the tenuous outer layers of our planet's gaseous envelope, and it is then that the fun starts.

By far the most common gases in Earth's atmosphere are nitrogen and oxygen and it is the interaction of atoms of these gases with the charged solar wind particles that give rise to the light and colors of the aurora. Atoms of nitrogen become ionized as electrons are stripped away, and atoms of both gases are raised to an excited state by incoming solar corpuscles as they are funneled down and accelerated along lines of magnetic field bunching together near the magnetic poles. When an ionized nitrogen atom regains an electron, a photon of blue light is emitted. If an atom of nitrogen returns to ground state from an excited state, a photon of red light is given off. Oxygen atoms returning from excited to ground states emit photons of either green or brownish-red, depending on how much energy had been absorbed.

Oxygen takes an unusually long time to return from an excited to ground state. It can take three quarters of a second to emit a green photon and up to as long as 2 min to emit a lower energy red one. During this time, the atom will probably collide with other atoms or molecules. The energy of these collisions drains away the excitation energy and stop photon emission. This is especially the case for the tardy red emissions. Nevertheless, at very high altitudes, the air (which, by the way, has a high percentage of oxygen at these levels) is so rarefied that collisions between atoms are infrequent enough for many to have time to emit red photons. At lower altitudes, red emissions seldom happen, but atomic crashes are nevertheless infrequent enough for a good percentage of excited oxygen atoms to emit green photons. At still lower levels in the atmosphere, atoms become so crowded that there is insufficient time between atomic collisions even for green photons to be emitted.

The result is a gradation of auroral colors with altitude. At the greatest altitudes, brownish-red predominates. Below this we find the green emissions of oxygen and the blue and red of nitrogen. At the lowest levels, green disappears, leaving only the blues and reds of nitrogen.

Oxygen green is the most common color in auroras, although if the display is faint, color may be difficult to discern. If there is a suspicious glow pole-ward of your location, a small hand-held spectroscope will determine if it is auroral by showing the green line of oxygen.

Mixing also gives rise to more colors. The frequently observed pink, for instance, is a mixture of light green and red. Red and blue likewise can mix together to give yellow. Pure blue is the least reported color in auroral displays.

As seen from outer space, auroral activity forms a ring around the Earth's magnetic poles. The radius of this ring is normally about 1,600 miles or 2,500 km. At times of greatest activity, the ring expands, with the auroral zones migrating south in the northern hemisphere and north in the southern. It is then that displays are seen at temperate latitudes. These migrations mark the times

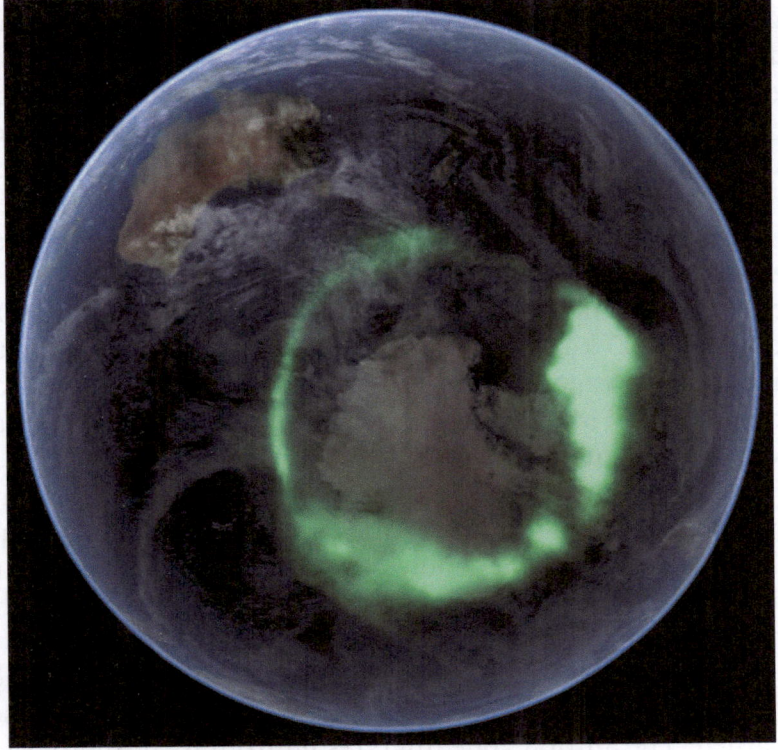

Fig. 3.10. NASA's IMAGE satellite captures the aurora Australia from space on 11 September, 2005 (Credit: NASA)

of magnetic storms, when the velocity of the solar wind and the strength of the interplanetary magnetic field both increase.

Conditions such as these are more likely to happen during the intense phase of the solar cycle, with the 3 years (or thereabouts) after the actual time of solar maximum being the most prolific. It is then that flares and generally heightened solar activity causes the solar wind to be most blustery.

For some reason, auroras happen more frequently around the time of the equinoxes. Why the position of Earth should be relevant to a phenomenon which has its origin in the Sun is not completely clear, although it is known that at these times the interplanetary magnetic field and the geomagnetic field link up. At the magnetopause, Earth's magnetic field points north and when the slant of the interplanetary magnetic field is large and tilting south, the Earth's field at this point of contact is partially cancelled out, effectively opening a window to the solar wind.

This happens during April and October, simply as a matter of geometry. Because the Sun is rotating, the interplanetary magnetic field lines are spiral instead of straight, hence the slant at Earth's magnetosphere. Moreover, because the Sun's axis of rotation is tilted at 8° to the plane of Earth's orbit, there are times of greater exposure to solar wind particles emitting from the Sun's pole and there are other times when Earth will be level with the solar equator. Interestingly, the solar wind streaming away from the solar poles has a higher velocity than that coming from lower solar latitudes; 1,012,500 miles (1,620,000 km) per hour as against 900,000 miles (1,440,000 km) for the average velocity. The times that our planet is more exposed to these greater velocities occur around September 5 and March 5; the dates that Earth reaches its greatest heliocentric latitude.

Although both of these situations work together to increase auroral tendency around these times, they do not appear adequate by themselves to tell the whole story. Something else appears to be at work here as well, but what that "something else" might be is anybody's guess at present.

There are times when the Sun appears to go manic, and on these occasions the Earth's magnetic field does not escape unscathed. The auroral zone migrates far from its normal location and beautiful light displays are witnessed by folk for whom the northern and southern lights are normally little more than tales from exotic climes. Less attractively, these occasions are also times of severe magnetic storms, even triggering large scale power outages.

Such was the great solar event of August 1, 2010 when much of the Earth-facing hemisphere of the Sun erupted into a frenzy of activity (see Fig. 3.7). A large flare appeared, a wave-like "solar tsunami" swept across a major region of the solar disk and multiple filaments streamed outward from the photosphere as the corona seemed to shake with activity. A coronal mass ejection poured outward into interplanetary space like a huge bubble of hot plasma. Shortly thereafter the disturbance reached Earth triggering auroral displays well outside the normal zones. Two days later, auroras were witnessed as far from the Pole as Wisconsin and Iowa in the United States, almost down to 40° northern latitude. During times such as these, dwellers in the Earth's temperate zones have a chance to witness scenes normally reserved only for folk living at high latitudes. Even so, the auroral displays seen at lower latitudes are not as spectacular as those closer to the true home of these nocturnal spectacles.

Extraterrestrial Auroras

Earth is not the only planet in the Solar System that experiences auroras. The gas giant worlds, Jupiter and Saturn, both have stronger magnetic fields than Earth (over 14 times stronger in Jupiter's case) and each possesses a large radiation belt. Not surprisingly, auroras have been observed near the magnetic poles of each, both from spacecraft passing near these worlds and with the Hubble Space Telescope nearer home. Similar auroras have also been seen on Uranus and Neptune, the "ice giants" of the outer Solar System.

Polar auroras on these planets, like those of Earth, are powered by the solar wind, but in Jupiter's case, some interesting auroras of a different origin have also been recorded. These are due to electric currents along field lines – so-called "field aligned currents" – generated by a dynamo mechanism springing from the relative motion of the rotating Jupiter and the orbital motion of its

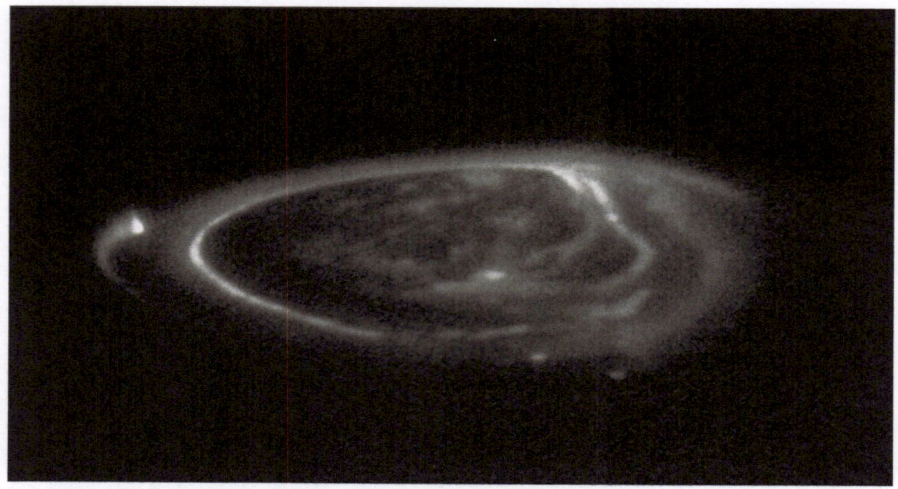

Fig. 3.11. Jupiter's auroral zone (Credit: NASA)

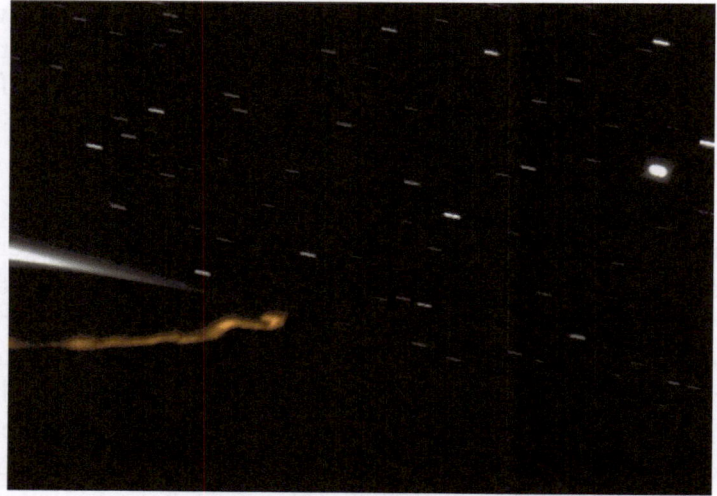

Fig. 3.12. Auroral activity high above the northern region of Saturn, as imaged by the Cassini spacecraft (Credit: NASA)

nearest large moon, Io. The presence of these currents was first detected as radio emissions back in 1955, when they were believed by some to be due to very energetic lightning. But if Io contributes to auroral activity on Jupiter, Jupiter returns the compliment!

The large Jovian moons Io, Europa and Ganymede have very thin atmospheres and the interaction between these tenuous envelopes and the magnetospheric plasma of the giant planet triggers auroral activity on all three of these satellites.

The smaller rocky planets Venus and Mars are also known to display aurora. In the case of Venus, there is no planetary magnetic field similar to that of Earth and its auroras are therefore not restricted to polar latitudes. They appear only as bright and diffuse patches which may even spread across the entire planetary disc. They are caused by electrons from the solar wind striking the planet's nocturnal atmosphere. Venusian auroral displays vary considerably in form and intensity and are less like their terrestrial counterparts than the solar-wind induced displays of the giant worlds.

Mars likewise has no global magnetic field of the type known on Earth, but data from the Mars Global Surveyor shows a complex of crustal magnetic anomalies and, above an area where the strongest localized magnetic field is found (a region known as Terra Cimmeria, at southern latitude 52° on the Martian globe) an aurora was noted by Mars Express' SPICAM instrument on August 14, 2004. The area of auroral emission was quite confined, measuring just under 19 miles (30 km) across and 5 miles (8 km) high. It is believed to have been caused by a flow of electrons along the crustal magnetic lines of force, exciting molecules in the planet's upper atmosphere.

With the discovery of planets beyond our own solar system, the prospect of extraterrestrial auroras takes on a whole new dimension. Many Jupiter-like planets and even super-jupiters have now been discovered, some hugging their parent stars in orbits far smaller than that of Mercury in our own solar system. Presumably these giants possess strong magnetic fields and, being so close to their parent suns, we can only imagine the sort of auroras that flash and shimmer around heir magnetic poles. Then, what can we say about planets that orbit red dwarf stars, of the type prone to flares that put to shame anything erupting from the Sun (and a good thing for us too, I might add!). What auroral displays might these super flares trigger in the atmospheres of orbiting worlds?

By cosmic standards, Earth's auroras might be quite feeble affairs!

Project 5: Observing Auroras

If you live at high northern latitudes, aurora will be common sights in the sky . (Unfortunately for southerners, the south magnetic pole is over the ocean, so the relatively few dwellers at far southern latitudes are not as fortunate as their northern counterparts). But if you are further from the magnetic pole, auroras will be more or less rare and every opportunity should be made to see them. Keep watch on "space weather" forecasts and when conditions are favorable, take careful note of what might be happening along your pole-ward horizon.

Typically, a display begins rather unspectacularly as a simple glow along the northern horizon (or southern, for southern hemisphere observers). In time, this may rise to become an arc. The lower part of the arc may brighten significantly and shoot forth multitudes of streamers, eventually losing its arc-like form and assuming the pattern of a billowing curtain of streamers. In a large aurora, these streamers may shoot past the zenith and converge in a spectacular corona as the display reaches its climax. During the declining phase, waves of light may pulsate from the horizon in a display known as "flaming."

Take note of the colors of the display, its duration and any unusual features which might be noted, including any of the anomalies mentioned in the next section of this chapter.

Aurora-Like Anomalies

Thus far, we have been looking at "normal" auroral phenomena, on Earth and elsewhere in the Solar System. But there exist on record reported observations of aurora-like events which are hard to explain in conventional terms. It is tempting to disregard these, but the history of science has taught us that this is not often a viable option. Scientific knowledge expands through encountering "anomalies"; observations that don't "fit" existing ideas and theories. If the anomaly won't go away, the only alternative is to revise our ideas and sometimes this has led to truly revolutionary advances in knowledge. There may be nothing revolutionary here, but even if there is not, heed should still be paid to the observations that "do not fit in." Indeed, the phenomenon of the aurora itself did not "fit" with known theory until the true nature of the atom was discovered, as we have already remarked. For all we know the explanation of the following anomalies might await another – even if less fundamental – advance in knowledge.

Black Aurora

Writing in the journal *Knowledge* in 1894, Mr. A. C. Ranyard describes a most unusual phenomenon in the northern skies of Croydon, England. What he witnessed was a pattern of "five dark rays or streamers" emanating from a point over the northern horizon. These "radiated like spokes of a wheel toward the zenith" and appeared clearly defined against a hazy sky faintly illuminated by the glow of city lights. The whole system appeared to rotate slowly through an arc of 30–40° toward the east before fading out. Ranyard described the rays as steady – they "did not seem to quiver or change color, but remained black or smoky, and reminded me of streaks made by a hair pencil dipped in Chinese ink."

Even stranger was the phenomenon observed on March 24, 1903, at 1.40 P.M. in broad daylight from Somerset in England. The witness, Edward Nelson, described a "black arch with streamers radiating upward from it; in short it was precisely like the commonest form of auroral display, as seen in high latitudes, only in negative, the streamers and arch being black instead of white, and the background blue sky instead of a night sky." Mr. Nelson said that the arch was located in the western part of the sky and added that the day was fine, with the Sun shining, but with some "bright white clouds ... driven in front of the aurora at a rapid rate by a fresh breeze."

This daylight observation is somewhat reminiscent (although probably not directly related to) various tales which surface from time to time about cloud formations – or what appear to be cloud formations – appearing in the afternoon and turning into auroras after sunset. Aurora are not normally considered bright enough to be seen in daylight, although the tops of some go high enough to project beyond the Earth's shadow on occasions. Nevertheless, these alleged "daylight aurora" are not said to be dark, unlike the phenomenon which Nelson here describes!

Low Level Auroras?

From what has been said earlier, it will be appreciated that the conditions favorable for the formation of auroral phenomena occur only at high altitude. Low level auroras, it would seem, are simply not possible. But the only trouble is, people have seen them!

For example, Mr. James Helvor Johnson, writing in the *Publications of the Astronomical Society of the Pacific*, recounted an experience at Eagle, Alaska, in the winter of 1901. Johnson witnessed a spectacular aurora whose streamers descended below the line of a relatively nearby 1200-ft bluff. He was quite sure that the aurora "was in a line between me and the bluff. The streamers were about one-quarter of the way down below the summit of the bluff, and no part thereof reached to the summit." Also interesting was Johnson's comment that "There was a swishing sound" as the auroral streamers moved. The issue of auroral sounds will be raised later in the present chapter.

Also in 1901, an aurora was witnessed in September and recounted many years later in an article by C. S. Beals in the *Journal of the Royal Astronomical Society of Canada*. The observer was on the south bank of the Yukon River at the time, in an area where the river "is about half a mile wide." He claims that the auroral streamers came between where he was located and the northern shore, right to the water's edge. He remarked that "Of this there was not the slightest doubt, as the trees [on the timbered northern river bank] could be seen through the openings in the auroral streamers."

Beals also, in the same article, referred to an account by Q. M. S. Griswold in 1925 who saw an auroral curtain "approach to within 4 ft of the surface of the Earth." A building could be seen "beyond and through the curtain of light" indicating that its proximity was not an illusion. Moreover, Griswold claimed that he actually walked right through the auroral curtain (!), saying that there was no sensation as he did so and that the curtain itself "disappeared from the view of anyone approaching closely; however; other people about 100 yards from the curtain could see the observer enter and pass through the curtain."

These accounts are far from being the only ones in the literature. Auroras have been reported beneath clouds, in front of ranges of mountains and on at least one occasion, even between the observer's eyes and his outstretched hand! Moreover, on several occasions, parallaxes measured using theodolites have indicated altitudes of various auroral features ranging from 4 miles to just 2,000 ft.

Observations such as these are "impossible" according to the accepted understanding of aurora. Yet, they have been so frequently reported by apparently credible witnesses, and even supported by a few instrumental measurements, that they are difficult to dismiss as mere illusions or mistakes. A skeptic will probably point to the relative age of most of the reports (the majority come from journals published during or prior to the first half of last century) as reason for doubt. However, observers then were no less cautious than we are today and there is no real reason to disbelieve older reports merely because they are older. Indeed, the opposite might well be true. Before the true nature of aurora was known, the sighting of auroral activity close to ground level may have seemed more credible and anyone witnessing this might have been less cautious about reporting it. Today, anyone seeing a ground-level aurora is more likely to keep quiet about it for fear of ridicule or, at best, being met with a wall of disbelief. They may not even believe it themselves. The eyes might say "I saw this", only to have the brain reply "You could not have seen it!" and the eyes yield.

Yet, after admitting this, it remains true that nobody really knows what to do with reports such as these. Explanations range from illusion and misidentification – which may truly account for some reports, but it is very difficult to squeeze all of them into this restrictive mould – to the occurrence of some type of low-altitude luminescence (possibly akin to St. Elmo's fire, which we shall meet in the next chapter) co-incident with, but not necessarily directly related to, an ordinary (high altitude) aurora occurring at the same time.

Is That an Aurora I Hear (and Smell!)?

Alternatively, is it possible that an aurora may sometimes trigger a sort of secondary electrical luminescence at or near ground level that may, from the point of view of a nearby observer, appear to merge with it into a continuous luminous phenomenon? Everyone knows how difficult it is to estimate distances, especially at night, and it is possible that two luminous phenomena at vastly diverse distances might mistakenly be judged as continuous. What we then would have is not strictly speaking a low-altitude aurora per se, but a different sort of electrical phenomenon which is both triggered by an aurora and which, at least superficially, appears to mimic its form.

It will be appreciated just how speculative this suggestion is, but there just might be some evidence supporting it. Evidence at least, supporting electrical activity at ground level during auroral events, although not directly associating this activity with any form of luminosity. This in itself is controversial, though studies since the late 1970s make it appear less bizarre today than it may have in earlier times.

The evidence of which I speak concerns the sound and odor of auroras. Yes, you read that correctly! Over the years a number of reports of sounds and smells associated with these displays have been published in reputable science journals.

For example, an entry in an early (1824) edition of the *American Journal of Science* mentions a resident of the Faroe islands who stated that "when the aurora borealis is dark red, and extends from west to east with a violent motion" an odor "similar to that which is perceived when an electric machine is in action" is noticeable. Much more recently, in 1941 at New Jersey, an ozone-like smell was reported by several people during an intense auroral display on September 18 of that year. Earlier, in 1870 in Norway, Rollier, the balloonist, reported an aurora which was both heard and smelt as well as being at low altitude. Quoting from the report published in *Sky and Telescope* for September 1946, the account reads "In 1870, Rollier ... descended on a mountain in Norway 1,300 m high, saw auroral rays across a thin mist, and heard a muttering. When the sound ceased he perceived a very strong smell of sulfur."

Other auroral aromas resembling "the smell of smoke and burnt salt" have also been reported at times.

It is interesting that these types of smells have often been said to accompany nearly lightning flashes and other forms of electrical activity, and it is in this respect that the reported auroral odors may be cited as evidence favoring some sort of electrical discharge in the immediate vicinity of the observers.

More frequently reported than auroral odors, auroral sounds open up a fascinating area of research which, until the last decades of the previous century, was considered by most scientists to lie beyond the pale of serious investigation. We have already seen how Rollier characterized a sound accompanying the auroral streamers witnessed during his balloon descent as "a muttering." Writing in *Nature* on May 19, 1881, Mr. F. C. Constable recalls previous correspondents to that journal as describing auroral sounds in terms of "crackling", "the flickering of blazing fire" and "rustling or switching of silk" and compares these descriptions with his own experience of sounds like the "constant crackling or rustling of blazing flames" heard during a magnificent display of sheet lightning.

Like Constable, many witnesses have reported much the same sounds from auroras, lightning flashes and bright meteors. Typical descriptions include "sizzling", "swishing", "popping", "crackling", "rustling", "hissing" and so forth. Reports of meteor sounds of these types have gone back to before the time of Edmond Halley and, for just as long, they have been dismissed by most scientists (including Halley, whose adventurous scientific speculations stopped short of meteor sounds!) as psychological illusions.

In fairness to the skeptics however, the biggest problem with these sounds was their timing with respect to the sighting of the meteor, lightning flash or aurora. Unlike thunder or the sonic booms of bright meteors, they were heard instantaneously with the visual sighting. But because sound is a lot slower than light, that should not happen. To arrive with the light, sound would need to be generated before the light, which in the examples of meteors and lightning strokes is, of course, impossible.

A few scientists did take the reports seriously however, but it was not until Colin Keay, Professor of Physics at the University of Newcastle (NSW) in Australia began to investigate the phenomena associated with a brilliant fireball seen in the Sydney/ Newcastle region of New South Wales in 1978 that the real break-through came. The account of this is given in my book *Weird Astronomy* and will be only dealt with briefly here. Essentially, what Keay found was that a portion of Earth's magnetic field gets trapped and twisted in the turbulent plasma trail of a meteor. As the plasma cools again and the ionization neutralizes itself, energy in the form of radio waves is released and these are picked up by objects in the observers' neighborhood and turned into sound. Keay found that even a person's hair or spectacles can act as antennae, explaining another curious fact concerning these strange sounds; some people hear them while others standing nearby do not. This fact had previously been raised as evidence for their subjective – psychological – nature, but all it really means is that a frizzy haired person or somewhere wearing spectacles has the advantage of an antenna whereas a bald-headed observer or some-one with naked eyes does not!

Once the reality of these *electrophonic* sounds was estab-lished, it became easier to accept other long-range electromagnetic effects as well. For instance, a bright meteor in 1992 was held responsible for an electric shock received by a motorist, thanks to an induced electric current in the frame of his automobile.

If a fleeting meteor is capable of inducing such electrical dis-turbances at ground level, is it so difficult to believe that a sus-tained and active auroral display could not only be accompanied by electrophonic sounds, but also trigger ground-level discharges sufficiently strong to give rise to odors and even to be observable as luminous displays in the lower atmosphere?

Incidentally, back in 1881–1882, an interesting and thus far unrepeated experiment was carried out by one Professor Selim Lemstrom in Lapland, that may have a bearing on the subject of low-altitude auroras. The professor mounted arrays of wires with sharp points atop insulated posts in several places, most notably on the top of Mt. Pietarintunturi, about 40 miles from Sodankyla. Although the experiment was only carried out there on 2 days, on the first day – December 29, 1882 – an auroral streamer formed directly over the top of Lemstrom's apparatus! If this was not a purely co-incidental formation of a normal streamer at typical alti-tude, this event could constitute further support for an interaction between auroras and electrical phenomena close to the ground. Unfortunately, no similar experiment has been performed in more recent years, so the possible implication of Lemstrom's work to the present discussion is uncertain.

Non-auroral Glows in the Night Sky

Although the most spectacular, the polar aurora is not the only source of nocturnal illumination. Two other sources with very different origins can be distinguished which, although never brilliant, are nonetheless interesting in their own right and contribute important information to our knowledge of Earth's cosmic neighborhood. We shall now look at these in turn.

A Glowing Sky

For an astronomer, that would seem about the gloomiest possible prospect! Yet, it is something that we have always lived with. From the Earth's surface, the sky really is luminous. Fortunately, only faintly so, but luminous nonetheless.

The first person to identify this faint, perpetual aurora-like phenomenon was the Swedish scientist Anders Angstrom, who identified it back in 1868. Combined with starlight, it is the reason why the night sky does not look completely black. Actually, it is surprising just how much light really does come from the sky when one is fully dark adapted and well clear of any artificial source of light. The only reason we are not more aware of this nocturnal glow is its relative uniformity over the whole sky.

Of course, it was known long before Angstrom that the sky, even on a moonless night, is not completely dark. Anyone who notices the silhouette of an outstretched hand against the night sky knows that. One suggestion even held that the light from the night sky must be due to the enormous number of faint meteors burning up in our atmosphere. But it was Angstrom who found that at least part of this glow is due to the presence of a green emission line in the spectrum. He found that this emission line, believed to be associated with auroral activity, is always faintly present in the night sky, even when there is no auroral activity.

But it was not until the 1920s that further progress in the understanding of the phenomenon was made. Early investigations were made by Robert Strutt, the fourth Baron Rayleigh and son of John Strutt, the third Baron Rayleigh of "Rayleigh scattering" fame. About the same time, John McLennon and G. M. Shrum identified the green spectral emission line as being due to atomic oxygen. Soon thereafter, V. M. Slipher discovered the sodium layer and its contribution to the glow of the night sky and in 1931, Sydney Chapman suggested that the glow is caused by the combination of molecules, high in the atmosphere, previously split apart by the action of sunlight. During the following years, the complexity of the emissions became more apparent and the term "airglow" was coined in 1950 as a blanket term covering the combined effect of these various emissions.

Airglow is now commonly divided into three main forms, dayglow, twilightglow and nightglow.

The first of these occurs while the atmosphere is being directly illuminated by the Sun and is intrinsically the brightest. It is caused by a combination of resonance and fluorescence. These processes occur when molecules are excited by the absorption of radiation from the Sun. In the first, light of the same frequency or color as the absorbed incident light is emitted. In the second, the emitted light is of a lower frequency or "redder" color than the incident light absorbed. Although this form of airglow is the brightest of all three, it is the most difficult to detect as it is overwhelmed by both direct and scattered sunlight.

Twilightglow occurs when only the upper atmosphere is directly illuminated by sunlight and is the most easily observed form of the phenomenon by observers on the ground. This is because the ground and lower atmosphere are in darkness, even as the upper atmosphere continues to be activated by direct sunlight.

Nightglow occurs later at night when the entire atmosphere is in darkness. It is fainter than the dayglow and twilightglow and the dominant process causing it is a different one. Nightglow is mainly the result of chemiluminescence. During the daylight hours, solar radiation breaks apart molecules which subsequently recombine at night, emitting photons of light in the process. The lion's share of the light is contributed by chemical reactions between oxygen and nitrogen atoms and molecules and hydroxyl molecules. These reactions take place in the upper mesosphere of

Fig. 3.13. Nocturnal airglow (nightglow) (Credit: Angelo Antonio Leithold)

Earth's atmosphere at altitudes of between 60 miles (100 km) and about 190 miles (300 km). Although this airglow is faint, the total amount of light given off still makes a greater contribution to nocturnal illumination than the combined illumination of all the stars.

Also contributing somewhat to the total light emission of the upper atmosphere are the processes of photodissociation (i.e. emission of light from the excited states of neutral fragments of molecules caused by solar radiation), photoionization (similar to the last, except that here the fragments are also ionized), inelastic collisions of high-energy electrons produced in the former process and a very minor contribution from excitation by cosmic rays from beyond the Solar System.

Although the airglow is normally diffused over the sky, it becomes modified by heating effects during magnetic storms to form broad and diffuse arcs aligned along geomagnetic parallels between about 40–55° of latitude and some 60–250 miles (300–400 km) altitude. These features, known as stable auroral red (or SAR) arcs were discovered in 1956 over France by Barbier. They show up brightly in the wavelength of red oxygen emission, but are not noticed by the human eye, whose sensitivity to that wavelength is very slight.

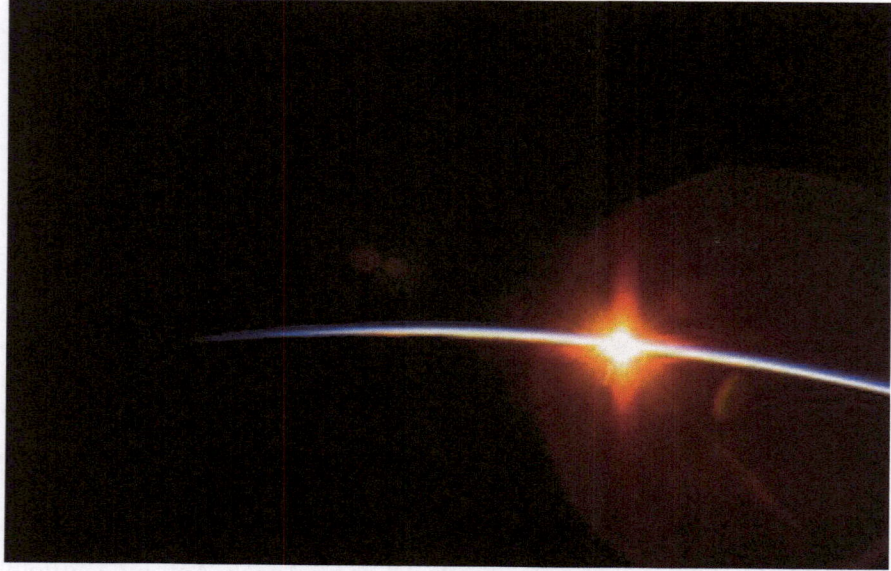

Fig. 3.14. Setting Sun and thin airglow layer as seen from space, 19 April 2006. This image was obtained by an Expedition 13 crewmember from the window of the International Space Station (Credit: NASA)

Project 6: Observing the Airglow

Although the airglow is normally diffused pretty evenly across the sky, the brightest region forms a rather thick zone about 10 miles (16 km or thereabouts) deep at an altitude of around 60 miles (100 km). If we look directly overhead, we see this zone broadside and its surface intensity is faint. However, if we turn our gaze more toward the horizon, our line of sight takes in increasingly greater depths of airglow. Looking very low over the horizon however, also means that we are peering through greater depths of atmosphere and this has the effect of absorbing the pale light of the airglow. As a compromise, look about 10° above the horizon (approximately two spans of an outstretched hand) on a very clear and dark night. Do you detect a pale glow all around the horizon at that altitude?

Airglows on Other Worlds

Earth is not the Solar System's only planet on which airglow has been observed. As early as the 1980s, the *International Ultraviolet Explorer* orbiting observatory picked up emissions from Venus caused by the combining of atoms of nitrogen and oxygen to form a molecule of nitric oxide. Later, the *Venus Express* spacecraft detected emission in the near infrared part of the spectrum resulting from the same process.

Likewise, *Mariners 6, 7* and *9* observed, at ultraviolet wavelengths, airglow on the dayside of Mars, however nothing was detected on the night side of that planet until 2004 when the European Space Agency's *Mars Explorer* orbiter discovered the elusive nocturnal glow in the atmosphere of Mars. The Martian atmosphere contains a small amount of molecular nitrogen and on its daytime side, some of these molecules are broken down into free nitrogen atoms, thanks to irradiation by extreme ultraviolet light from the Sun. Molecules of water vapor and, especially, the ubiquitous carbon dioxide are also broken down by solar ultraviolet, resulting in a small amount of atomic oxygen in the Martian atmosphere. Molecular oxygen is first released from molecules of carbon dioxide, but these are broken down still further into the atomic state.

After the Sun sets on the Martian scene, we might expect the nitrogen and oxygen atoms to combine to form nitric oxide, however the upper atmosphere of Mars, where most of this photodissociation takes place, is so thin that the chances of atoms coming into contact with one another is very small. First of all, the atoms of nitrogen and oxygen must move downward toward the Martian surface to an altitude where increased atmospheric density makes recombination possible. It is only then that they combine into nitric oxide molecules, emitting the characteristic flashes of light as they do so. This Martian airglow layer is located 37–50 miles (60–80 km) above the planet's surface.

Discovery of the airglow layer showed that air was moving downward from high altitudes toward the surface of Mars. But this was not happening all over the planet. It was happening only at high southern latitudes during the extreme cold and darkness of a southern Martian winter. The detection of airglow on Mars thereby became valuable evidence for sinking air over the south pole of the planet during winter time!

Banded Airglows

Although the airglow is normally a diffuse and featureless luminescence, distinct patterns of banded light are reported from time
to time. These bands are usually parallel, but if extending across
much of the sky appear to diverge and converge due to the effect of
perspective – the "railway line effect."

Although it is possible that some reports of "banded skies"
are caused by nothing more unusual than thin streaks of cirrus
clouds, it is thought that the genuine cases result from the passage
of acoustic gravity waves through the upper atmosphere at altitudes of around 50–60 miles (80–100 km). In effect, the bands are
"ripples" in the airglow caused by the passage of these waves.

The mention of "gravity waves" conjures up thoughts of ripples in space and time triggered by orbiting neutron stars and collapsing (or colliding!) black holes or some such exotic event
occurring in deep space. But these are not the types of gravity
waves relevant here. Acoustic gravity waves are far less exotic.
They can best be thought of as oscillations of wind and temperature that are accompanied by little or no change in pressure. Also
called "buoyancy waves", they can occur between any stable layers of fluids of different density. The ripples spreading out on the
surface of a pond after a stone is dropped into the water are gravity
waves. Analogous waves in the atmosphere can be triggered by
such events as auroral disturbances taking place in the ionosphere
and even by meteorological phenomenon such as towering thunderstorms that punch high into the troposphere, or by cyclones
and kinks in the jet streams. As well as causing bands in the airglow, acoustic gravity waves also cause ripple patterns of clouds,
although not all cloudy ripples are due to these waves.

Interesting displays of banded airglows were observed by
Australian astronomer and dedicated weather watcher Gordon
Garradd in 1989 December 22, 26, 28 and 1990 January 23.
Garradd noted that the display on the latter date was particularly
intense; the bands having an apparent surface brightness similar to
that of the Large Magellanic Cloud.

Garradd remarked that "When passing overhead the bands
moved by up to 20° *toward* the South over a 10 min interval; which
indicates that the source of the gravity waves was apparently

Fig. 3.15. Satellite image of gravity waves over the Arabian Sea, 23 May 2005 (Credit: NASA-GSFC)

not auroral. My notes from 1989 Dec 22 ... say that the bands of light of East–west orientation were first noticed at 11 UT, and that there was a general glow in the West SW and South (probably auroral)." He further notes that "On some occasions there were a dozen or more bands across the whole of the sky" and that the bands overhead had disappeared by 13 UT, although the bands then visible in both the north and the south were "quite bright" in the sense that they were comparable with the pyramid of the zodiacal light. As we will see in the following section, the zodiacal light is conspicuous in dark rural skies, but melts into the background light pollution near large urban centers. These displays of banded skies, like the airglow itself, are not for the city dweller!

In the early 2000s, *bores* were found to occur in the upper mesosphere and to leave their signature on the airglow. Defined as horizontally-propagating jumps in the density, pressure or temperature, bores have long been recognized as a common phenomenon in tidal estuaries, as well as in the ocean and lower atmosphere. More recently they have been discovered at much higher altitudes. Mesopheric bores can be either *undular* – where the leading front is followed by a series of trailing waves, phase-locked to the initial wave – or *turbulent*, where the leading front is followed by a turbulent region. All-sky images of the airglow – or even naked eyes under good conditions – reveal these features as the bores march across the sky.

Zodiacal Light

More conspicuous than the airglow but less noticeable than the aurora, the *zodiacal light* can be seen under any clear and dark sky following evening and preceding morning twilight. In appearance, it takes the form of a cone or pyramid of light, broadest at the base and tapering off as it reaches higher into the sky. The base of the cone stretches along the horizon for at least 30° and the cone itself may be traced half way to the zenith. It appears over the region of the horizon where the Sun has set in the evening or where it will later rise in the morning, extending upwards along the zodiac as it wanes in intensity. The morning cone has long been known as the "false dawn"; a term found alike in early Islamic literature and more modern literature, such as the "westerns" of Zane Grey. Because of the better orientation of the zodiac as seen from tropical climes, the zodiacal light (in contrast to the aurora) is best observed from the tropics. In the northern hemisphere, the evening (western) cone is best seen in late winter and early spring and the morning (eastern) cone from October to December. The reverse is true for the southern hemisphere.

Although the cones wane in intensity as they reach higher into the night sky, under very good conditions they are seen to merge into a very faint band of light stretching right across the sky and joining the evening and the morning cones. Estimates of the width of this zodiacal band range from 5 to as much as 20°. Mid way along this band, a slight brightening may be seen in the form of a faint oval patch some 10° wide. Although the total brightness

Fig. 3.16. Zodiacal Light cone as seen from Paranal, 19 November, 2009 (Credit: ESO/Y. Belatsky)

of this patch is around that of the star Sirius, it is so diffused that the surface intensity is very low and a really dark sky is required to see it. Interestingly, when astronomer E. E. Barnard discovered it late in the nineteenth century, he initially thought that he had found a large and extremely diffuse comet passing very close to Earth!

The zodiacal light differs from both aurora and airglow in so far as it is located out in space, well beyond Earth's atmosphere. It also differs from these phenomena in not being self luminous. Its light is merely reflected sunlight; but reflected from what?

The answer is dust – interplanetary dust. We now know that the Sun possesses a disk of dust spread through the inner planetary system and it is this that we see as the three aspects of the zodiacal light phenomenon. Because the particles are small, the phenomenon of forward scattering of sunlight plays an important role in making the zodiacal dust cloud visible. Just as wind blown spider webs and balls of thistle down "light up" when they pass in front of the Sun on a clear day, so cosmic dust particles similarly appear to shine brightly when the Sun/particle/Earth angle is large; between approximately 110° and 180°. The effect is also seen in dusty comets that pass between Earth and Sun at these large angles. It is possible for the apparent brightness of such an object to increase 10,000-fold under these circumstances, causing it to blaze out in full daylight almost on the solar limb!

Thanks to this forward scattering effect, the zodiacal dust becomes relatively brightly illuminated at these large Sun/particle/Earth angles (known more accurately as "phase angles"). This is responsible for the zodiacal cones in the west after evening twilight and in the east before dawn. But the extension along the entire length of the zodiac is much fainter because the phase angles are no longer large and forward scattering ceases to occur. However, opposite the Sun, where the phase angle approaches zero degrees, another (albeit weaker) scattering phenomenon known as "back scattering" enters the scene and causes the region of the zodiacal band at very *small* phase angles to intensify again, though it never becomes as intense as the cones themselves. It is this effect that is responsible for the phenomenon known as the *gegenschein* or *counter glow* – the faint oval light patch mentioned above. This can be thought of, in effect, as a pale reflection of the

Fig. 3.17. The pale glow of the gegenschein (Credit: Jimmy Westlake (Colorado Mountain College))

Sun on interplanetary dust. Because each reflecting particle is opposite the Sun in the sky – as viewed from Earth – it is in "full" phase in the sense that the particle's Sun-lit hemisphere is turned fully toward us. It is the combined effect of all these miniscule "full moons" scattering reflected sunlight back toward their source; in the process, giving us the gegenschein.

The gegenschein was not always recognized as a back-scattering effect however. Its constant anti-solar direction led a few astronomers to propose a different explanation in the form of a terrestrial tail. The Earth, they supposed, sported a tail like that of a comet! Although this would have been interesting and was quite an ingenious explanation for the phenomenon, it became untenable once spacecraft far from Earth also sited the gegenschein in *their* anti-solar direction. Clearly, like the other phenomena associated with the zodiacal light, the cause of the gegenschein lies far beyond our world.

Project 7: The Zodiacal Light

If skies at your location are clear and dark, the brightest section of the zodiacal light should be relatively easy to see in the form of cones of light – broader at the base and tapering off as they rise further into the sky – in the western sky after twilight and in the east before the first fingers of dawn. Residents of large cities are definitely at a disadvantage here, but those in rural settings will find the cones easy to observe and might like to plot them on full-sky star charts, noting how their positions and orientation varies during the course of a year.

For observers in really dark locations (where stars of magnitude 6 are relatively easily seen with the naked eye near the zenith) should be able to trace the fainter zodiacal band right across the sky, joining the morning and evening cones. Directly opposite the Sun, the third component of the zodiacal light – the gegenschein or counter glow – presents an interesting challenge. Wait until the region of the gegenschein is high in the sky (towards midnight), stay in the dark for at least 20 min and then carefully examine the region with the naked eye. Far from artificial lights and under skies where stars of magnitude 7 are glimpsed by the eye alone, it is surprising how "conspicuous" this large, diffuse, glow can be! How easily can you see it?

Lest it be thought that the plane of the planetary orbits is choked with dust, it should be mentioned that the zodiacal cloud is really very tenuous by terrestrial standards. It has been estimated that if the dust grains average 1 mm in diameter and have the same reflecting power as the surface of the Moon, their average separation would be about 5 miles or 8 km. In actual fact, sizes and reflectivity will vary, but the general idea is clear. This is hardly a cosmic sandstorm!

Although we might think that comets supply most of the dust particles comprising the zodiacal cloud, it now seems that asteroids are also major contributors. Comets certainly puff out a lot of dust, but the most conspicuous cometary dust formations (the curving tails of bright comets) consist of very fine particles that are quickly swept right out of the Solar System and therefore make no contribution to the zodiacal cloud. Moreover, long-period comets show scant regard for the plane of the planetary orbits, so even the larger particles which they leave behind generally spend little time in the region of the zodiacal cloud. Comets moving in orbits of small inclination, especially those of short period, make some contribution, but much of the dust which finds its way into the cloud started its journey by being kicked up from asteroid surfaces by impacting meteorites. These impacts range from relatively insignificant strikes to the rare catastrophic collisions between two asteroids. Since the mid-1990s, the phenomenon of Main Belt comets or MBCs – asteroids which periodically or sporadically go through phases as active comets – has been recognized, and some of the zodiacal dust must also be supplied by these objects. We do not as yet know how many MBCs there are, so the importance of their dust contribution remains unclear. Probably, there are not many active at any one time, but we know that at least two of the six known at the time of writing currently become active at each perihelion passage (there has not yet been time to track the others around their orbits), so the dust contributions of at least some of these are quite regular. We also do not know how many asteroids that are inert at present might have been active in the past or may become active in the future. And, of course, just because the MBC's are faint, it is almost certainly the case that we have as yet only seen the tip of the iceberg of those that *are* currently active.

Since the IRAS infrared survey in 1983, dust bands are known to be associated with several families main-belt asteroids. These "families" almost certainly consist of fragments of asteroids smashed by mutual collisions millions of years ago. Because of the numbers of family members perusing similar orbits, secondary collisions between members must also occur. Indeed, direct evidence of this comes in the form of sub-families within larger groups. Dust raised by these continuing impacts explains the dust bands associated with them. Interestingly, the Themis family of asteroids is one group possessing a prominent dust band, and it is also the one with which three of the recently discovered MBC's are associated. This may be no co-incidence.

If asteroids and comets (Main Belt and otherwise) did not keep up a constant supply of dust, the zodiacal cloud would slowly disappear and this source of nocturnal illumination would be no more. As they orbit the Sun, the dust particles of the zodiacal cloud absorb solar energy and re-emit it again in all directions. This simply means that they become warm and radiate "warmth" or infrared radiation, but radiation itself exerts pressure. Sunlight exerts pressure on the tiny particles of a comet's dust tail and drives them off into interstellar space. Similarly, the radiation emitting from the somewhat larger particles in orbit around the Sun exerts pressure on the particle itself. Radiation emitted from the particle in the direction of its motion exerts a pressure slowing the particle and causing it to gradually drop into a smaller and smaller orbit until, eventually, it will spiral down into the Sun. This is known as the Poynting-Robertson effect, and through its operation the zodiacal cloud is being continually eroded away. Or it would be, were it not for the continuous supply of new particles being shed by asteroids and comets. As the ancient Greek sage Heraclites once said of a river, the zodiacal cloud is never the same twice. Fresh water (or in this instance, fresh dust) is ever flowing in as the old drains away.

Apparently a balance is maintained at this time, but maybe the intensity of the zodiacal light fluctuates over the ages and there may have been times in the past when it glowed far more brilliantly than it does today. A major asteroid collision or the deflection of a giant comet into an orbit of short period presumably results in significant ejections of dust into the zodiacal cloud and dramatic intensifications of the light. Astronomical historians Victor Clube and Bill Napier, whom we shall meet at greater length in the following chapter, argue that the zodiacal light was much more intense just a few thousand years ago, tied in that instance to the activity of a giant comet that became the progenitor of the modern Comet Encke and the Taurid meteor complex. Their position, as we shall suggest, is not convincing in its finer details, but their wider point of view concerning the evolution of cometary debris complexes is worthy of further consideration and it seems entirely logical that the present condition of the Solar System with respect to interplanetary debris need not remain constant throughout time. Secular fluctuations – perhaps major

Fig. 3.18. Comet 2P/Encke, 5 January 1994 photographed by Jim Scotti using the Spacewatch telescope at Kitt Peak (Credit: NASA)

ones – in the intensity of the zodiacal light would be expected as a consequence of these changing conditions. Maybe the skies under which our remote ancestors lived – and our distant descendents will live – were/will be illuminated by far more dramatic zodiacal light phenomena than we know today. Now that's an interesting thought on which to ponder as we contemplate these ghostly nocturnal glows!

4. Lightning's Gleaming Rod

I saw the lightning's gleaming rod
Reach forth and write upon the sky
The awful autograph of God

Joaquin Miller

An active thunderstorm is one of the most awe-inspiring and spectacular sights that we can witness. What pyrotechnic display can compete with it? What laser show can be its equal? If it were not for the fact that, for most of us, such displays are a relatively frequent occurrence, they would be held in far greater esteem – and (dare I say it?) – fear than is actually the case. Imagine if thunderstorms were so rare that the average human being could expect to experience one only once or twice in a lifetime. How would they then be regarded?

Nature's Static Electricity Generator

Thunderstorms generate electricity at a fantastic rate, driven by the rapidly rising currents of air that give thunderclouds their towering form. Thanks to these powerful vertical winds, positive and negative charges are separated and enormous potentials built up between the upper and lower regions of the cloud as well as between cloud base and ground. However, the exact mechanisms by which these charges are separated is not at all clear. In fact, several processes could be at work and may well be operating in tandem.

Ice appears to be an important player in the thunderstorm's dynamo. Even in the summertime, when the strong upward currents of air building thunderclouds are driven by the heating of the ground, the great cumulus clouds tower to such altitudes that their tops poke into freezing layers of air. It is the predominance of ice crystals in the cloud's upper levels that give the mature

D.A.J. Seargent, *Weird Weather: Tales of Astronomical and Atmospheric Anomalies*, Astronomers' Universe, DOI 10.1007/978-1-4614-3070-4_4,
© Springer Science+Business Media New York 2012

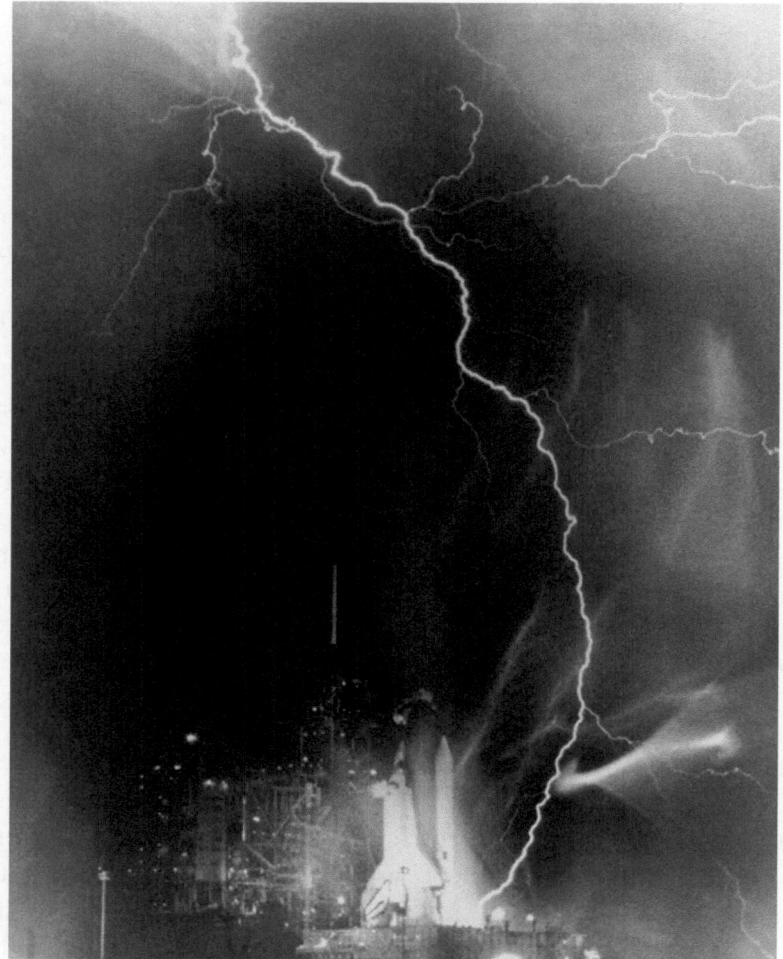

Fig. 4.1. Lightning creates an eerie light display in the hours preceding launch of Space Shuttle Challenger, 30 August, 1982 (Credit: NASA)

thundercloud, or cumulonimbus, its characteristic anvil top as high altitude winds waft away the ice crystals into great sheets of cirrus cloud, spreading away from the summit of the cumulonimbus like smoke from some gigantic chimney stack. But freezing does more than give thunderclouds their anvil top. Experiments have shown that if very small amounts of impurities are present, significant differences in electrical potential exist between ice and liquid water when the two are in contact. When water is allowed to drip onto a block of ice, a charge difference results if the water is very slightly impure. Most of the impurities in rainwater cause the ice to become negative and the liquid water positive. In a typical thunderstorm, heavier ice particles forming high in the cloud fall earthward, eventually entering regions of higher temperature where they tend to collect drops of liquid water very close to the freezing point. Some of these cold, but still liquid, droplets will partially turn to ice upon contact with the falling pellets. The segment of a drop in contact with an ice pellet will freeze while the rest of the drop splashes away. As in the laboratory, the pellet becomes negatively and the "splashed away" water positively charged.

Having acquired an extra load of ice, the pellet falls even faster into still lower regions of the cloud while the positively charged water is carried higher in the developing storm's updrafts. In this way, charges are separated in the cloud – the upper regions becoming positive and the lower negative – until electrical potentials become so large that lightning starts flashing.

Fig. 4.2. Lightning over Launch Pad 39A, Kennedy Space Center, August 25, 2009 (Credit: Justin Deniere/EPA)

Other electrical generation mechanisms obviously exist however, as lightning is also sometimes seen in clouds that do not reach freezing levels. In these cases, capture of atmospheric ions by falling raindrops may play the dominating role.

Whatever the exact sequences of processes in any particular storm cell, the end result is truly amazing. Differences in electrical potential amounting to many hundreds of millions of volts are built up between the upper and lower regions of the thundercloud. To place this in perspective, a potential difference of 32,000 V will drive a spark 1 cm (less than half an inch) through air. Yet lighting bolts can be miles long; a truly awesome voltage! And the popular expression "quick as lightning" is not without merit either. A bolt of lightning can travel at the amazing speed of 130,000 miles per hour. Incidentally, within this gigantic electric spark, air is heated to temperatures approaching 20,000°C (36,000°F) or around three times the heat of the surface of the Sun. In every respect, lightning is a phenomenon destined to fill us with awe and wonder.

A typical cloud-to-ground lightning flash begins with a negatively charged channel of ionized air known as a *leader* proceeding down from the cloud base toward the ground. Almost invisible to the eye, it proceeds in steps toward the ground, taking the relatively long time (by lightning standards!) of hundreds of milliseconds to travel the distance. Beneath it, the electric field on the ground is enhanced and becomes strongest on emergent objects such as tree tops, towers and other tall buildings. Hence the warning never to take shelter under a tree during a thunderstorm. If the field grows strong enough, positive discharges called *streamers* snake upwards from these points and if one of these connects with a downward leader, an ionized channel opens between cloud and ground allowing a far greater current to return from the ground to the cloud. This is called the return stroke and is the brightest and most energetic part of a lightning discharge. The majority of lightning strikes are made up of at least four re-strikes along the same discharge channel. Air heated by the lightning stroke compresses cooler surrounding air to create a supersonic shock wave, which subsequently decays into the acoustic wave we know as thunder.

At any instant, there are some 6,000–8,000 thunderstorms occurring somewhere in the world, contributing somewhere in the order of 100 lightning flashes each second or over eight and a half

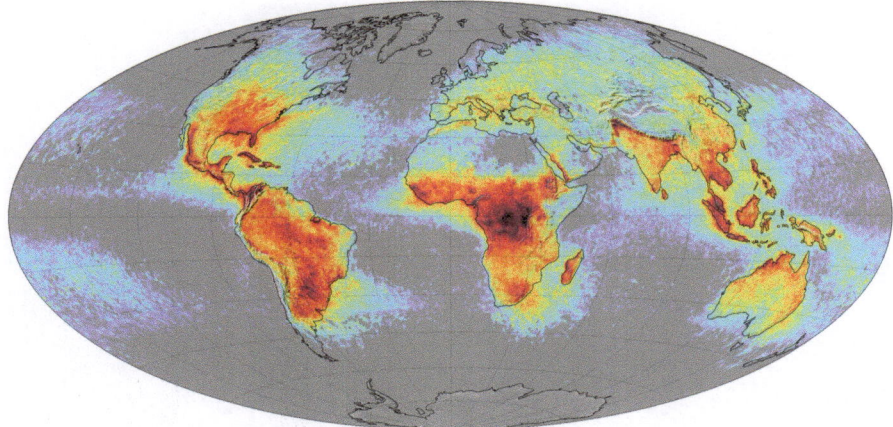

Fig. 4.3. Patterns of lightning frequency from data collected by NASA satellites between 1995 and 2002 (Credit: NASA)

million each and every day. Each stroke sends negative charges into the ground (with some less frequent *positive* exceptions, as we shall see later) as equivalent positive charges flow upward into the ionosphere. These upward discharges, long known in theory only, have relatively recently been observed as a strange ghostly double of lightning, as we shall see in due course.

Types of Lightning

Lightning is informally subdivided into chain, fork, sheet and heat lightning. These terms are, however, not the officially recognized ones. Chain lightning is only another name for streak lightning, especially for those flashes where are definite zigzag pattern is noticeable. Similarly, fork lightning (or for*ked* lightning) is the popular name given to a cloud-to-ground discharge in which several branches are apparent. Sheet lightning is simply the reflection of lightning hidden by cloud (mainly cloud-to-cloud discharges) and heat lightning, traditionally playing around the horizon on hot summer evenings (but, at least in the writer's part of the world, also fairly common on the ocean horizon in winter when an off-shore wind is blowing) is simply very distant lightning,

Fig. 4.4. Cloud-to ground lightning (Credit: US National Oceanic and Atmospheric Administration)

sometimes so far away that the clouds producing it lie out of sight beyond the horizon.

More formally, the following varieties are distinguished:

Streak lightning. This term refers to the bright return stroke of a lightning flash and is the one that occurs most frequently, although not the most commonly seen as many flashes remain hidden in the clouds.

Cloud-to-ground lightning. This is the one most commonly seen and is the second most frequent type to occur. It is also the most dangerous, for obvious reasons. These are the flashes initiated by a leader stroke progressing downward from the base of a cloud, normally a cumulonimbus or "thundercloud."

Staccato lightning. This is a sub-type of the previous variety, distinguishing itself as a single and very bright flash which frequently has considerable branching of the "forked lightning" appearance.

Ribbon lightning. We noted earlier that return strokes normally use the channel created by the original discharge. However, if a thunderstorm is accompanied by strong cross winds, this channel moves and each successive return stroke becomes slightly displaced from the previous one, giving rise to a ribbon-like pattern of discharges.

Ground-to-cloud lightning. Like cloud-to-ground lightning, this form of discharge bridges the ground and base of a thundercloud, however in this instance, the lightning discharge is initiated by a negatively-charged leader moving upward from the ground and meeting positively-charged ions from the cloud. The return stroke goes back to the ground from the thundercloud. This form of lightning is a lot rarer than the typical cloud-to-ground flash.

Cloud-to-cloud lightning. This is simply lightning that does not come down to the ground. It occurs in two varieties; inter-cloud lightning or flashes between clouds and the more common intra-cloud lightning that discharges between different regions of the same thundercloud. Frequently, this latter variety plays through the spreading anvils of well developed or decaying thunderstorms where it may divide spectacularly into a multiplicity of branching strokes. Such flashes are sometimes given the name of *Anvil Crawlers* for obvious reasons.

Positive lightning. Sometimes an unusually powerful flash of lightning will come from the thunderstorm anvil and strike the ground well outside the thunderstorm's periphery. For this reason, they are also called *bolts from the blue* as they seem to strike from a relatively clear sky. The leader initiating these strokes travels several miles in a horizontal direction from the positively-charged top of the thundercloud before turning earthward and meeting a negatively-charged streamer rising from the ground. These giant flashes typically carry between six and ten times the voltage of a normal (negative) flash and last around ten times longer. Fortunately, positive lightning flashes only constitute around 5% of the total.

Fig. 4.5. Intra-cloud lightning (Credit: US National Oceanic and Atmospheric Administration)

Peculiar Lightning

All of the abovementioned types are more or less "regular" forms of lightning. True, the last entry in the list is relatively rare (which, considering its incredibly high voltage and propensity to come to earth far from the storm centre, is undoubtedly a good thing for us!) but it is still a well documented variety.

Yet, in addition to assuming these "regular" forms, lightning can also manifest in some very peculiar guises.

Beads of Light

First, and probably closest to the "regular" varieties, is *bead lightning*, also known as *pearl lightning*, *segmented lightning*, *punctuated lightning* or, sometimes, *chain lightning*. (The last epithet is a little confusing, as the same term is sometimes used as a synonym for the familiar streak lightning, especially in instances where many "zigzags" are visible.)

This relatively rare variety occurs when an otherwise normal cloud-to-ground flash appears to break up into a chain of small and bright condensations that fade out more slowly than the original flash. Why this occurs remains poorly understood. One plausible sounding explanation is that, in these flashes, the width of the original lightning channel varies along its length and, as it cools immediately following the electrical discharge, the wider sections remain hotter for longer and so remain luminous after the narrower sections fade out. This appears to fit the descriptions of the string of beads effect, but it immediately asks why some lightning flashes have variable widths while most do not.

On September 2, 2010, a spectacular bolt of lightning in Adelaide, South Australia, was captured by camera and a slow-motion replay of the event presented on the weather segment of that evening's national television news. The growth of the lightning from a relatively faint streak to a brilliant band that saturated the image was well shown and there may even have been a suggestion of a pinch effect (see below) at maximum intensity, although the saturation of the image made it very difficult to be sure. What was also interesting was the way the lightning channel decayed. As it faded, in the very last minute fraction of a second, it broke into two or three small segments which persisted for just an instant after the rest of the lightning stroke had gone. A full string of beads effect did not occur, and the broken segments were too small and their duration too brief to observe except for the slow motion of the film, but it raises the question as to whether some minor and very brief "beading" of dying lightning flashes might be more common than we suspect.

The pinch effect is a possible source of constrictions in lightning bolts and may possibly be capable of disrupting the bolt into a string of luminous sections at times. *Pinched lightning* has been observed and photographed, although the classic photograph of this phenomenon (taken at Los Alamos, New Mexico in August 1961) is so unusual that it has even been reproduced in UFO literature as an alien spacecraft taking off! This flash consisted of one long and quite fat cigar-shaped segment, two shorter ones of similar appearance, a third very small and possibly fragmented blob and a fourth faint spot separated from the others at the end of a very faint string of light! No wonder some folk thought it looked like a UFO, but the segmented appearance also bears a similarity – albeit an exaggerated one – to the more "familiar" bead lightning.

The pinch effect is known in plasma physics and occurs when the magnetic field of a powerful current within a plasma causes a constriction of that plasma. It is widely present in nature, which is not too surprising as most of the matter in the universe exists in the form of plasma! The presence of pinch effects has been raised in phenomena as diverse as auroral streamers, solar flares, stellar eruptions and even quasars, all of which at first sight seem a far cry from bead lightning. Interestingly however, a study plotting the fluctuating current from lightning striking Mt. San Salvatore revealed a startling similarity between the resultant graph and that of the varying brightness of Nova Herculis of 1934! In each instance, the energy level (current in one instance and luminosity in the other) rose rapidly to a peak, then fell rapidly only to rise again more slowly to a second peak. The second peak in the lightning current was apparently due to release of pressure that had built up due to the pinch effect. Although similarity of graphs does not necessarily mean similarity of cause, simply finding a matching pattern in such phenomena of vastly differing magnitude is very interesting indeed.

The following account from South Africa is pretty typical of bead lightning reports.

> After the flash [striking the ground a short distance away] had died away, there remained a string of bright luminous beads in the path of the flash. The beads, of which there were twenty or thirty, appeared to be about a quarter of the width of the flash, that is, say, three inches in diameter. The distance between the beads, which appeared to be nearly constant, seemed about two feet. They remained visible for approximately half a second; during this time they gave no indication of any movement.

Flashes as Slow as Skyrockets

Another rather odd form of lightning is the so-called *rocket lightning*; a form of cloud-to-cloud discharge (mostly horizontal and at the cloud base) where the luminous channel advances so slowly that the eye can clearly follow its progress. Frequently, the forward advance of this type of lightning is intermittent, proceeding in small steps so to speak.

The term "rocket lightning" is also sometimes applied to (generally "slow") straight flashes that travel upward from the top of thunderclouds and into clear air.

Great (and Little) Balls of Fire!

Most controversial of all however, is the strange and difficult-to-explain phenomenon of *ball lightning* also at times known as *fire-balls*; somewhat confusingly insofar as this term is more typically given to bright meteors. For a long time many meteorologists dismissed the very existence of this phenomenon, explaining away reports as being due to everything from after-images to outright fabrications. Some probably doubt its existence to this day, although the weight of evidence for its occurrence is now too great for casual dismissal.

Personally, I have good reason to accept its reality. Back in the mid-1970s, I was watching a lighting display several miles distant when a bright cloud-to-ground flash left in its wake a brilliant greenish colored ball of light not far above ground level. At the time, I was reminded of the luminous ball from a "Roman Candle" firework, albeit obviously larger because of the greater distance at which I was seeing it. The "ball" remained stationary for a second or thereabouts before fading away. Several years later, I was informed of a very similar incident, except that this time the "ball" appeared a lot higher in the sky.

These accounts involve a simple type of ball lightning. Other reports point to something more complex; more dynamic.

Staying, for the time being, more or less within the writer's personal experience, an incident witnessed a few years earlier than the abovementioned ones was probably evidence of a more energetic and low-altitude form of ball lighting. I was visiting a close relative in hospital while a thunderstorm flashed and rumbled outside. All of a sudden, there was a single loud report – more reminiscent of the discharge of a double-barreled shotgun than a thunderclap – followed immediately by a loud crash of thunder. Although I did not personally see anything, a second person in the

Fig. 4.6. Ball lightning, as depicted by a nineteenth century engraving

room later said that a "bright object" flew past the window immediately before the loud report. The "bright object" was not, however, like a flash of streak lightning. Apparently, it appeared more as a discrete body than a ribbon of light. Perhaps it is worth mentioning that a brick factory with several tall chimney stacks was located next door to the hospital, close to where this object was sighted.

Another incident involving apparent ball lightning was reported by a family known to me. On this occasion, the mother and teenage son were inside a house while a thunderstorm raged outside. The son was lying back in an easy chair reading, with his legs and bare feet stretched out before him. Suddenly, a luminous ball appeared inside the house and rushed down the hallway. The son yelped as it singed the ends of his bare toes, before either leaving the house or disappearing.

With these personal and semi-personal experiences, I can hardly deny the existence of ball lightning. Yet, I am less confident that all reports attributed to it can be accommodated under the heading of a single phenomenon. At the very least, if ball lighting is the name for a single phenomenon, it is certainly a much diversified one. The above accounts alone show quite a range, but a better indication of just how wide this range truly is, becomes more apparent by taking a look at a cross section of the reports accumulated over the course of many years.

Thus, what are we to make of the following report from Pennsylvania in the year 1898?

> In July, 1898, ... as the sun shone brilliantly from a cloudless sky [note this! – DS], there appeared in front of the writer, apparently not more than 100 feet away and 50 feet elevated from the earth passing from an easterly to a westerly direction, three balls of candescent light linked together, descending like a bolt of lightning and accompanied with the most terrifying and hissing sound, ending in a report much like the sudden immersion of a large molten mass of iron into a body of water.

Similarly, in Dresden on May 12, 1912, a witness saw two reddish balls of light, one somewhat larger than the other, float through the air connected by what seemed to be a luminous thread. In this instance, there was a thunderstorm relatively nearby, although the balls of light did not appear to be closely associated with it.

Ball lighting also varies greatly in size.

For instance, a curious record dating from 1876 tells how two ladies at Ringstead Bay in England witnessed a truly amazing display of luminous balls near a cliff top. Apparently, there was no thunderstorm in the area at the time. The report stated that the luminous objects were "the size of billiard balls" and "[surrounded the ladies] on all sides." It continued,

> The balls were all aglow, but not dazzling, with a soft, superb iridescence, rich and warm of hue ... Their numbers were continually fluctuating; at times thousands of them enveloped the observers, and a few minutes afterwards the numbers would dwindle to perhaps as few as twenty, but soon they would be swarming again as numerous as ever.

The report also mentioned that the little globes of light continually eluded the grasp of the ladies; at times coming almost up to them, only to retreat again once the ladies attempted to tough them.

Another example of miniature ball lightning was observed to attach itself to the finger (!) of a certain Mrs. Ames of Baltimore, Maryland, on June 19, 1924. This occurred during a thunderstorm, but the lady in question was inside a house at the time when a rather faint oval ball of light "the size of a pecan [nut]" rose from the floor and attached to her finger. Shortly thereafter "There came a flood of lightning outside" and the ball disappeared.

At the other end of the scale of sizes, a lightning ball seen at Dyfed in West Wales on June 8, 1977 was described as "about the size of a bus"!

Similarly, just 2 years earlier in Albany, New York, a globe of ball lightning possibly 15 ft, or thereabouts, in diameter appeared in conjunction with a lightning bolt of unusually long duration that struck a factory building. And long before either of these events, back in 1930, a witness in Nebraska told of "A lavender colored shapeless mass" floating downward at the approach of a squall line. At the same time "globular structures, 28 and 42 ft in diameter" ran along the power lines. Incidentally, the greenish luminous ball that I recounted seeing earlier must have been pretty large, to have appeared as large and bright as it looked to me at such distance.

Although the term "ball" normally conjures up the idea of something more or less spherical, some examples of ball lightning have assumed a markedly different shape. For example, a witness named Lillian Mack, living in Kansas City, was at home during a thunderstorm one afternoon in 1959 when she heard what sounded like "crashing glass." Immediately following this noise, "An object about 2 ft long and 1 in. in diameter came flying like a spent arrow into the room where it hovered while forming the shape of a ball. After half a minute, it dissolved." While dissolving, a sound like breaking glass was again heard, yet a search of the house afterwards failed to find any glass that had been broken. Presumably, the "breaking glass" sound was, on each occasion, related more to crackling electrical discharges than to smashing windows.

Years earlier a small group of men in Burlington, Vermont, reported seeing a rocket-like object "about 6 ft long by 8 in. in diameter" sail through the air "about 50 ft above the tops of the buildings" following a sound which one of their number described as "[the noise of] a most unusual and terrific explosion." Although this "explosion" was "evidently very near by", no damage or any other indication that an explosion had occurred was found. At the time, the weather was fine, although a large thundercloud was approaching the location of the event.

The line of demarcation between reports of low-level nocturnal lights (we will meet these in Chap. 7) and ball lightning becomes blurry in the most unusual cases. For instance, although an account from Rhodesia (now Zimbabwe) in 1910 of a transparent "ball of fire, ... about 1 ft 6 in. by 10 in. in size" and casting only feeble light is included in a *Knowledge* article on "Globular Lightning," the differences between this low energy ball of light and the energetic globes more typical of ball lighting casts doubt as to whether the two are all that closely related. Moreover, this account states that the light "remained in one position for about 5 min" allowing the witness to come "within three yards" of it before "moving away in a zigzag fashion at a rate of about 6 miles an hour", finally entering a belt of trees where it is said to have disappeared (though whether this "disappearance" was simply becoming hidden amongst the trees or truly fading out is not stated in the report). No mention is made of weather conditions at the time but because the omission of any reference to thunder or lighting would be curious if this had been present, it might be safe to assume that the event did not happen during a thunderstorm.

Then there are some curious reports of "dark" ball lightning. Early photographs of ordinary streak lightning sometimes give the impression that the lighting bolts are black, however this is merely a photographic freak known as film reversal. The black streaks on photographs are really just negative images of ordinary bright ones. Yet, this explanation cannot account for the rare but seemingly quite strong observations of dark ball lighting, all of which have been made visually. Witnesses of this phenomenon speak of "balls of smoke" following lightning bolts. However, the phenomena appears to be more energetic than just glorified smoke rings as at least some of the accounts also mention "sparks" being emitted by the balls and one very old account – from December 1596 – even speaks of a dark ball "exploding" upon entering a church building. This rare phenomenon is not easily explained, but the suggestion that the smoke (if that is really what it is) might be caused by the incineration of combustible material somehow drawn into the lightning ball during its formation seems a plausible one.

Ball lightning in general remains a mystery. Gone are the days (fortunately!) when it was dismissed as at worst a hoax and urban legend and, at best, an instance of after-images on the retina following a brilliant flash of lighting. These "explanations" were just insulting to the witnesses. Most people now accept that it is real, even if its explanation is still far from clear. Small luminous balls of plasma have been known to appear in the presence of high-current DC equipment and similar phenomena have been deliberately produced in laboratories. Whether ball lighting is analogous to these plasmoids is, however, not clear. If the term "ball lightning" is an umbrella covering several visually similar but physically different phenomena (which is entirely possible) a one-theory-fits-all approach will most probably fail. Some reports of apparently low-energy ball lightning sound, as already remarked upon, suspiciously akin to those of ground level nocturnal lights. These, as noted, will be discussed in Chap. 7, but if we may be allowed to jump ahead for a moment, we just mention that some (and I stress *"some"*) of these appear explicable in terms of swarms or colonies of luminous insects. We may legitimately wonder if, just possibly, this class of "ball lightning" is amenable to a similar explanation. Maybe the insects were stirred up by the thundery conditions. But of course, this explanation (if it happens to be valid for any accounts!) does not even begin to explain the brilliant and obviously energetic varieties of the phenomenon. Whatever might be the explanation for the glowing globe that I saw suspended in a lightning channel after the flash had faded, it was certainly *not* a swarm of insects! Yet the Rhodesian report referred to earlier *may* – just possibly – be explicable in this way.

Slow Lightning

Normally a flash of lightning is one of the most fleeting of events, lasting for only a small fraction of a second. Yet, from time to time, flashes variously described as "slow" or "prolonged" are reported. Not to be confused with slowly propagating rocket lightning, these flashes usually travel at normal lightning speed, but then they appear to "freeze" or "hang suspended" in the sky for intervals of a second or longer. Personally, I think "frozen lightning" or "hanging lightning" is a better term than "slow lightning." It better describes the visual appearance of the flashes and is less likely to be confused with rocket lightning.

Whatever we may like to call it, this form of lightning is even further from explanation than ball lightning and, like the latter, is open to the "explanation" of after images.

My most memorable encounter with this type of lightning happened quite a number of years ago during one of the most spectacular thunderstorms that I have ever seen. The storm was very rich in intra-cloud lightning, much of which was hidden from direct sight but made its presence known in glorious illuminations of the cloud base. These moved across the sky in waves like the pulsations of an aurora viewed at high speed. Yet the most peculiar feature of the storm was the relatively large number of streak lightning flashes that appeared to freeze in situ. A flash would occur and then just hang their motionless in the sky like a photograph of a lightning bolt. It was as if a film of a lightning display had been suddenly stopped!

The interval that the lightning bolts hung motionless and frozen seemed at least 1 s; a very long time for a lightning flash. Maybe the subjective experience of time exaggerated their duration somewhat, but the length of visibility of these "slow" flashes compared with the normal ones was nevertheless (no pun intended!) striking.

Speaking to a friend, who had also witnessed the display and many of its long-duration flashes, the following day I was told that the duration of these strokes was not real. My friend took the view that the phenomenon was caused by after-images and was not a feature of the lightning strokes themselves.

I did not believe that then and I do not believe it now!

For one thing, although some of the slow flashes were truly brilliant, not all the brilliant ones (produced by that storm and by many others than I have seen) were slow. Had they been after-images, why was this effect not witnessed after every very bright flash?

Moreover, not all of the slow flashes were bright. One that I especially recall occurred after the storm had passed into the distance. This flash was (from my location) relatively faint, yet it froze there in the sky for just as long a duration as any of the brilliant ones. I cannot accept that it left an after-image on my retina. It was simply too faint!

Lastly, I was not alone when I saw these flashes. Yet the flashes that appeared to me to freeze, were also the ones that my companions saw to do likewise. And the normal short duration ones were likewise seen as such by my companions. If the phenomenon was literally and physically in the eye of the beholder, it seems strange to me that there should be such agreement amongst several "beholders."

So if slow lightning flashes are not merely after-images, how can they be explained?

Unfortunately, like many mysteries of the natural world, it is easier to say how they are not explained than it is to come up with the correct explanation!

We have already remarked that positive lightning flashes from the very tops of thunderclouds last longer than normal negative strokes, but even these are not normally as prolonged as slow lightning. In any case, the flashes that I witnessed during the abovementioned storm were clearly *not* positive bolts from the cloud tops.

We also noted that in conditions of strong cross winds, the lightning channel may drift laterally resulting in several successive strokes of ribbon lightning. If one viewed this phenomenon at such an angle that one's line of sight lay precisely along the direction of drift, each discharge would appear to be in exactly the same place as the previous one and, in theory at least, the ribbon effect should be replaced by the appearance of a single repeating stroke which may be perceived as a stroke of longer than normal duration. For this to work as an explanation of slow lightning however, each stroke would need to be exactly the same shape or the stroke would seem to develop odd loops and duplications. Of course, the chance of seeing all the lightning ribbons lined up in this way is miniscule and the probability of having several strokes in the same storm align with one's line of sight is too remote for serious consideration. And that does not even mention the fact that (in the abovementioned storm) sightings of slow lightning were made by well scattered observers. Parallax alone crashes this "explanation" before it even gets off the ground.

Although it might not appear obvious at first glance, slow lightning has a number of similarities with bead lightning. The latter is also "slow" in so far as it persists for anything from half a second to as long as 2 s after the original flash has disappeared. It also marks out the track of the original flash, "hanging frozen" along the lightning path in a manner reminiscent of slow lightning. Indeed, it is tempting to categorize bead lightning as a form of slow lightning. Viewed in this way, the principal difference between these two forms is simply that in bead lightning, the persistent luminous path breaks up into discrete sections while in (non-bead) slow lightning it does not.

Project 8: Slow Lightning

If you are fortunate enough to experience a storm with many "slow" flashes, an interesting experiment would simply be to video the lightning display and see whether some of the recorded flashes linger. A video record of "slow" flashes should put to rest any thought that these are simply retinal afterimages and careful examination of the images may also reveal any tendency for the flashes to break up into "beads." A disruption of the flashes into beads or discrete sections might not be discernible with the eye alone.

Some support for the association between these two types of lightning may come from two observations of bead lightning.

The first dates back to the early part of the nineteenth century. After stating that "several flashes left beaded trails" and that "the time during which the beads remained visible [was estimated to have been] at least 1 s", the witness then made the interesting remark that "It appeared to me that the *entire course of the flash remained luminous, with a dull red glow,* but that at intervals along the path bright points like sparks appeared to remain suspended in the air." [Italics mine].

The second observation was made on July 27, 1936 from Clarendon in Virginia. After a short description of a spectacular series of lightning bolts, the witness writes that "One of these [bolts], which seemed almost to hang in the sky only a few miles away, appeared to ignite something high in its course, leaving wisps of flame which persisted momentarily, detaching themselves from the stream and drifting eastward from the path of the discharge as if carried by the wind ... A portion of the path of the discharge ... was also indicated by a train of sparks persisting momentarily after the streak had gone."

The "train of sparks" seems to have been bead lightning, although there were some peculiarities insofar as a number of the "beads" drifted away from the lightning track. But note the witness' description of the original bolt seeming to "hang in the sky"; the exact phrase that came most naturally to me in describing the slow lightning flashes that I observed. This case appears to be an instance of a slow flash which further evolved into a string of beads, possibly suggesting a link between the two.

It is also interesting that this witness also described some of the lightning bolts associated with this storm as "appearing like ribbons." Indeed, the one that both "hung" and broke up into segments was of that form. This implied that they were rather wider than average, a description also given to a slow flash seen over the North Atlantic in 1956. Here, the witness described a "Slow, unusually wide flash of lightning" that seemed to hit a cloud, from which a normal lightning flash then discharged. It would be interesting to observe whether slow flashes have a tendency to appear wider than regular ones or whether these two examples are just co-incidental. All I can add is that the slow flashes that I have seen appeared to be quite normal in width. Indeed, they looked normal in every way except for their prolonged duration.

As a final thought, we mentioned in our brief look at bead lightning that the plasma pinch effect has been suggested as a means of obtaining the relatively persistent "beads" observed in this type of lightning. The pinch effect, which is produced by the magnetic field induced by an electric current flowing through a plasma, is one way of containing a plasma and has been studied in conjunction with the quest for controlled thermonuclear fusion. Containment is, of course, the big hurdle to viable controlled fusion reactions. Is it possible that, under certain circumstances, the pinch effect is capable of "containing" a ribbon of plasma along the entire length of a lightning discharge's path, at least for a second or thereabouts?

Silent Lightning

We normally think of thunder and lightning as inseparable. Thunder is simply the sound of lightning and it is probably only because of the gap between seeing the flash and hearing the sound that we continue to speak about these as if they were separate phenomena. Why do we say "I can hear distant thunder" instead of "I can hear distant lightning"? After all, what we are hearing simply *is* (the sound of) distant lightning. We don't have a separate name for (say) the sound of an explosion, so why do we have one for the sound of lightning?

Be that as it may, we expect lightning to make a noise. That is, we expect it to be accompanied by thunder. But that is not always the case.

Of course, most "silent" displays of lightning seen at night are simply associated with storms too far away for the accompanying thunder to be heard. More rarely, anomalous sound propagation appears to be the culprit. One instance, reported in the literature, involved a relatively nearby thunderstorm which appeared to be discharging completely silent lightning strokes. Later investigation however found that it was typically noisy according to residents of a town over which it passed, raising suggestions of anomalous sound propagation rather than anomalous lightning as the real cause.

Yet, at other times it does appear that lightning really does do its work in silence. At times, this lightning is anomalous in other respects as well. A classic instance was a spectacular and unusual display observed by one Dr. Knoche in Paraguay on October 3, 1927. This storm produced a magnificent display of streak and bead lightning, as well as "various anomalous phenomena ... including ... rapidly-moving orange-colored discharges, which ... resembled cylindrical masses of glowing gas; flashes that revolved like pinwheels; and ... hundreds of luminous arcs crowded together near the zenith, so dazzling in their brilliancy that [Dr. Knoch] had to close his eyes." Yet, all of this continued during "A ghastly quiet" which lasted for several hours. Later however, thunder was heard and developed into a continuous rumbling.

Another (almost) silent lightning display was recorded in *Nature* in 1899. Taking place in Shanghai, the witness to this reported that "the lightning played over nearly the whole of the exposed sky, sometimes six or seven streamers at a time lighting up the sky." He described these "streamers" as looking "more like the discharges in a vacuum tube" than normal forked lightning and commented that the thunder following even the closest and brightest flashes was nothing more than a dull rumble, similar to that of ordinary thunder heard at a great distance.

Flashes that do not reach the ground are typically less noisy than those that do, even though they can be very long and brilliant. Apparently, some are simply not sufficiently energetic to cause the usual booming noise of typical thunder.

Recalling a personal experience here, I recall many years ago watching a developing rain shower only a few miles distant and falling from a rather large cumulus congestus cloud. As the shower developed, a radio in the house gave several sporadic cracks of static, indicative of lightning discharges. This, I should mention, was in the days before AM radio had been replaced by FM and the family radio set could still double as a good lightning detector.

The bursts of static were loud and only began as the cumulus cloud developed, so there could be little doubt that they were caused by discharges within the cloud. Yet, despite the proximity of the developing shower, no lightning was seen nor thunder heard. Presumably, the discharges causing the static were faint and feeble, but would presumably have been seen illuminating the cloud at night in a display of silent lightning. Or would they still have been unseen; an example of "lightning" that was not only silent, but invisible as well?

Like a Bolt Out of the Blue

Lightning from an apparently clear sky – bolts from the blue – are a rare phenomenon, but in most instances they are not especially mysterious. As mentioned earlier, positive lightning from the spreading anvils of thunderstorms can travel large distances through clear air and strike well away from the core of the storm. Such bolts as these may appear to come from a clear sky and may strike even when the Sun is brightly shining. Yet, a little visual search of the sky will quickly reveal their true origin. A large menacing thundercloud!

On rarer occasions however, bolts of lightning have been reported from a genuinely clear sky.

For instance, the following report relates to two topographic map makers in the Missoula National forest of Montana on July 2, 1927.

As they were making one of these observations [relevant to their map making] toward a point southeast of Gold Peak both men saw a flash of lightning strike the ground almost on their line of alidade sight, and about 15 miles from them. This flash was followed by four others within the next few minutes. The first strike started a forest fire, the others did not. The phenomenon was most peculiar because all of these strikes descended almost vertically, apparently out of a blue sky, the nearest clouds being about 15 and 25 miles, respectively, from the area struck.

The report continues by saying that the bolts struck within an area of "not over half a mile in diameter." They were not between the two clouds (the type of cloud was, unfortunately, not given. Were they thunderclouds? We don't know.) The account concludes by stating that no thunder was heard from the flashes and no further lightning was seen.

Non-thunderstorm Lightning

Lightning is also a frequent accompaniment of volcanic eruptions, where it flashes menacingly through the already menacing plumes of ash. Although we have (fortunately!) witnessed no such event in historical times, no doubt a giant meteorite blasting its crater in our planet's crust would add to an already horrendous spectacle by generating flashes of lightning through its debris cloud.

Whirlwind Lightning?

Dust storms can also produce lightning and there is some anecdotal evidence that it is occasionally generated even by large dust *devils.* At least, that is the conclusion drawn from an alleged eye-witness account dating back to 1902 and published in a *Science* article some 28 years later. This observation is such a strange one that it needs repeating in some detail. According to the witness,

> I was on the White Mountain Apache reservation, in Arizona, on the day in question, when an ordinary desert whirlwind whirled into view from just around a southerly projecting point of land north of White River from the now-abandoned Fort Apache. I had just crossed a flat area among the hills where an ancient lava flow once spread out, forming a "lava lake," an area probably six miles across from the afore-mentioned point to the mountainous hills to the northwest, up which I was then ascending. In a moment [the whirlwind] began to gain momentum on entering the level country and in a minute more it was a roaring funnel that was hurling immense quantities of dirt and sand skyward so they formed an umbrella-like cloud around the apex of the whirling center. As the twister was coming in my direction, I shifted southward over a gulch to another ridge to escape its fury. On it came. It entered the canyon in which I had been only a minute before. Here as the canyon both wedged-in and ascended toward the mountains in the direction it was going, the rushing whirl became "angry" as it were. The day had been perfectly clear. Yet in a moment there were chain lightning and ripping thunder on every side, while at the same time the whirler uprooted trees and tore large-sized boulders from their places on the canyon walls, finally destroying itself in the canyon. From my observations I am inclined to believe that the electrical display that accompanied this whirl was due to the friction caused by the whirling debris.

Although this account apparently stands alone as an observation of dust-devil lightning, experiments have established the presence of significant electrical fields associated with these whirlwinds, probably (as the witness of the above-quoted account suggests) due to friction of the whirling dust, sand and other debris. From the above description, it would appear that this particular whirlwind became unusually severe, probably causing an abnormally large electrical potential to build up within the vortex. The electric fields accompanying dust devils of more typical strength simply do not become sufficiently intense for lightning to discharge.

This raises an interesting question. If large dust devils on Earth occasionally generate lightning, is it possible that some of the mega-dust devils observed on *Mars* also generate electrical discharges of some form?

Let me say immediately that there is (as yet) no evidence for any such thing, but then the evidence for terrestrial dust devil lightning is pretty thin as well. Some of the dust devils seen on Mars were really huge and Martian orbiters have photographed tracks in the dust that appear to have been formed by whirlwinds of truly tornadic intensity. One might imagine that static electric charges could build up to high levels within these whirling funnels and that some type of electrical discharge may, at least on rare occasions, eventuate.

Volcanic Lightning

Back on Earth, volcano-generated lightning has long been known. Broadly speaking, three types of volcano lightning have been distinguished. The first type, occurring during large eruptions is quite normal in appearance, except that it flashes through clouds

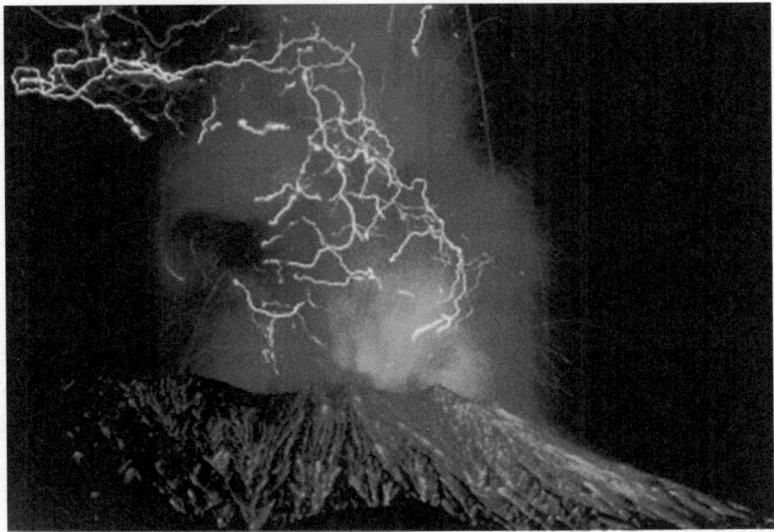

Fig. 4.7. Sakurajima volcanic lightning, 18 May, 1991 (Credit: Sakurajima Volcanological Observatory)

of volcanic ash instead of through ordinary thunderclouds. These lighting displays are sometimes called "dirty thunderstorms" for obvious reasons.

A second type of volcanic lightning is sometimes seen playing around volcanic vents. These strokes can shoot up from the active vent to an altitude of about 1.8 miles (3 km) above the volcano.

Both of these forms occur when water droplets and ice crystals interact with the electrically charged plume of volcanic ash.

The third type had been suspected, but not fully confirmed until the Redoubt Volcano in Alaska burst into action in early 2009.

Whilst monitoring an earlier eruption of the Augustine Volcano in 2006, Steve McNutt of the Alaska Volcano Observatory observed short flashes, just 1 m (about 3 ft) long and lasting for only a few milliseconds. These observations were unprecedented and appeared to represent a hitherto unrecorded phenomenon, so the chance to confirm them at Redoubt in 2009 was too good to miss.

McNutt and his colleagues set up a battery of equipment in late January 2009 in time for the big show in March when the volcano blew its top and, in so doing, turned on a spectacular display of "ordinary" lightning. To McNutt's delight, it also repeated the Augustine performance of miniature flashes.

Although McNutt's work has now established the existence of this miniature volcanic lightning, it is not clear how it is generated. One suggestion is that it is caused by electrically charged silica within the volcanic magma interacting with the atmosphere as it bursts out of the vent.

However it is caused, this hobbit-sized lightning may be a valuable discovery, according to University of Florida lightning expert Martin Uman. Uman suggests that as soon as a precursor event (a first warning of an impending eruption) is noted, instruments could be set up near the vents with the purpose of detecting the tiny lightning flashes. These seem to occur as soon as an eruption begins, and their detection would allow for early warnings to be sent out to surrounding areas as well as to air traffic.

This last type of volcano lightning also inclines us to be less skeptical about the following accounts, which on the face of it seem just a little too strange to be credible.

Miniature Thunderclouds and Allied Oddities

Thunderclouds are, of course, relatively large atmospheric features. As we have seen, the processes giving rise to lightning flashes are many and varied and it is difficult to imagine these taking place in radically scaled down circumstances.

What then, are we to make of "miniature thunderstorms"; Lilliputian thunderclouds whose dimensions are given in feet or meters rather than miles or kilometers? Do such things exist? Are they even possible?

It must be admitted that the evidence is not strong. Not, at least, if we confine ourselves to documented reports alone. But this writer at least has reason to suspect that the phenomenon might occur more often than these reports seem suggest at face value. I suspect that incidents have gone unreported either for fear of ridicule or simply from the assumption by some witnesses that "they [the experts on these matters] must know about it already." The reason I suspect this will be apparent in a little while, but first, let's look at what seems to be the best documented report and the one on which the weight of evidence rests.

The report was published by Mr. C. S. Bailey in the journal *Weather* under the title "A Miniature Thunderstorm." Although the account itself was not published until 1949, the event itself took place some 30 years earlier. As recounted by Bailey;

> I was staying at Stockton Heath [England] in July, about 400 yards from the Manchester Ship Canal. The evening was somewhat oppressive, and the air had become strangely still. Gazing down the road, I saw a small black thundercloud gathering along the length of the canal, and about 30 or 40 feet above it. It was approximately 400 yards long and 6 feet thick. As I gazed at this strange formation, a dazzling lightning flash raced through the entire cloud, i.e. parallel to the water, and a bang like the discharge of field artillery followed immediately. About 40 seconds later, another flash and report occurred; then the cloud thinned and dispersed in about four minutes. I might add, that at least in those days an air current of varying intensity moved up that canal almost incessantly, i.e. inland towards Manchester, at the Stockton Heath section, one felt it on the neighboring bridge.

This reads as a very straightforward account of a truly curious phenomenon. The only question I would like to ask is why he referred to the small cloud or fog layer as "a small black thundercloud." Was there something about its appearance that suggested this description or was it simply the events that followed? My thought is the latter, as Bailey's description suggests something more closely resembling a miniature stratus cloud than a scaled-down cumulonimbus.

An earlier and very vague account of something that may or may not have been similar is also found under the title "Curiosities of Thunderstorms" in an 1856 issue the *Electric Magazine*. In this article, reference is made to a statement by an Acamedician Marcolle, stating that a woman was killed by a flash of lightning emanating from a cloudlet just 1½ feet in diameter! No place or date is given for this alleged event and the "cloudlet" is not further described. It has been suggested that the "cloudlet" may really have been ball lightning, which if true, would place this event within a different category of phenomena.

Earlier, I mentioned that I had reason for thinking that this phenomenon, although certainly rare, might have been witnessed by more people than the very scant written records suggest. My reason for thinking this comes from a chance meeting about 20 years ago with a young man whom I did not know and have not seen since, but who seemed to be honest and open and who almost certainly knew nothing of the above accounts.

We were both waiting, separately, to speak with a business proprietor and as often happens in situations such as this, simply passed the time in idle conversation about all manner of things, not really focusing on any particular topic. Sometime during the conversation he casually mentioned that he liked thunderstorms and was a bit of a "dare devil" in this respect, enjoying time out of doors when the air fairly crackled with electricity. It was in connection with this that he recounted the following incident.

Apparently, he liked surfing and 1 day was out riding the waves with a group of likeminded friends when a sea fog rolled in. Now, unlike the Californian coast, sea fogs are relatively rare on the New South Wales seaboard where this incident took place, but what happened next was nothing less than uncanny. Some little distance from the group of surfers, tiny lightning-like flashes began discharging from the fog into the ocean. Apparently, this phenomenon was confined to a small region, although no mention was made about any difference in the fog at that point (whether it was unusually thick, for instance). The young man – typical of his approach – wanted to paddle his surfboard over for a closer look, but his companions managed to talk him out of it! Nothing more was said about how long the phenomenon persisted or how many flashes were seen, although I suspect from his general description that the discharges were quite numerous.

Superficially at least, this seems to be a very similar phenomenon to Mr. Bailey's miniature thunderstorm.

Explaining these strange reports is not easy. It is true that electrical potential increases significantly during foggy conditions. In fair weather, the potential between the head and feet of an average-sized person standing upright is somewhere in the order of 200 V. In foggy weather, this may increase to around 2,000 V. But even this is hardly enough to trigger lightning. It pales in significance compared with the 18,000 V, or thereabouts, that is reached between our head and feet during a desert dust storm (where lightning may be present). It seems that, somehow, the electrical potential can be magnified in certain small regions of the fog, but how this happens and by what mechanism these miniature lightning discharges are generated is a mystery.

Though probably not classifiable as a miniature thunderstorm, an account of a curious display of miniature lightning was reported by Brian Matthews in the journal *Weather* in 1964. The event took place on March 3 at Tucson, Arizona, during what was described as a heavy snowfall of large and wet flakes. During the snowstorm, Matthews noted that "Short flashes of 'lightning' were seen occurring at intervals estimated to be 15–20 s and at random places around the town." He described the flashes as single, short and without the usual "flicker" of normal lightning. They were also less intense than normal flashes (being too weak to cast shadows) and caused no static on the radio. Moreover, no thunder was heard during the display. From a vantage point atop an 80-ft observation tower, Matthews noted that the flashes were observed to originate "from points at or very close to the ground" and that "the illumination seemed to emanate from a single point which lit up the falling snow and cloud." Although this display of miniature lightning was not confined to a small region, its description sounds very similar to the discharges in the sea fog described above. The silence of the Matthews event, and the lack of any mention of "thunder" in the sea fog incident, is probably the chief difference to the Bailey observation, where loud reports accompanied both flashes. These flashes were evidently more energetic than the later examples.

The Eerie Blue Glow

Since ancient times, a strange bluish or violet luminescence has been reported, during thunderstorms, mainly around tall and/or pointed objects such as the masts of ships and even the horns of cattle. Because of its apparent liking for the masts of sailing ships, it came to be widely known as *St. Elmo's fire* in honor of St. Elmo or St. Erasmus, the patron saint of sailors. Other names such as *spirit candles, candles of the Holy Ghost* or *candles of St. David* were given to the phenomenon by Welsh sailors, while the Chinese knew of it as the *fire of Mazu*, from the Chinese sea-goddess of that name who was said to create the fire on ship's masts in order to guide and bless lost sailors. To the Portuguese, the strange lights were known as corposants or *corpusants;* a term derived from "corpo santo" or "holy body" while Russian sailors referred to them as *Saint Nicholas' lights, St. Peter's lights, St. Helen's fire* or *St. Hermes' fire.*

During his time aboard the *Beagle*, Charles Darwin writes, "we witnessed a splendid scene of natural fireworks; the masthead and yard-arm ends shone with St. Elmo's light; and the form of the vane could almost be traced, as if it had been rubbed with phosphorus."

James Braid, surgeon at Lord Hopetoun's mines in Lanarkshire, who encountered the phenomenon whilst on horseback on the night of February 20, 1817, graphically describes the appearance of the "fire." In his own words,

> It was about nine o'clock, P.M. I had no sooner got on horseback than I observed the tips of both the horse's ears to be quite luminous: the edges of my hat had the same appearance. I was soon deprived of these luminaries by a shower of moist snow which immediately began to fall. The horse's ears soon

became wet and lost their luminous appearance; but the edges of my hat, being longer of getting wet, continued to give the luminous appearance somewhat longer.

I could observe an immense number of minute sparks darting towards the horse's ears and the margin of my hat, which produced a very beautiful appearance, and I was sorry to be so soon deprived of it. The atmosphere in this neighborhood appeared to be very highly electrified for eight or ten days about this time. Thunder was heard occasionally from 15th. to 23rd, during which time the weather was very unsteady: frequent showers of hail, snow, rain &c.

From this close-up description (and the rim of one's hat is about as close up as one can get!), the "fire" was perceived as a display of tiny sparks. Indeed, St. Elmo's fire is not fire at all. It is plasma. A strong electric field surrounding the illuminated object results in the ionization of molecules of air creating the relatively faint discharges. The potential of the electric field required for the phenomenon varies between 100 and 3,000 kV/m, the strength in any particular instance depending greatly upon the shape of the object itself. Sharply pointed objects tend to become sites of St. Elmo's fire at lower potentials than objects that are not as pointed. This is because electric fields are more concentrated in areas of high curvature, such as sharp points, causing discharges to be more intense there. This is why lightning conductors are pointed; and also why it is a good reason to stay clear of pointed objects during a thunderstorm!

The blue or violet shade of the glowing plasma is caused by fluorescence of atmospheric nitrogen and oxygen. In effect, St. Elmo's fire is a natural "neon" light, except that it is powered by nitrogen and oxygen instead of neon. The fluorescence mechanism is also akin to that of aurora in Earth's upper atmosphere and in the atmospheres of other planets such as Jupiter and Saturn, as well as the glow of the gaseous component of the heads and tails of comets and of gaseous nebula in deep space.

St. Elmo's fire is generally associated with thunderstorms, but it may occur in other places as well. Volcanic eruptions provide another source of the phenomenon. It was observed on the masts and rigging of ships in the Sunda Strait near the violently erupting Krakatau volcano in 1883. This is not surprising considering the electrical activity that frequently accompanies volcanic eruptions in general and the exceptional Krakatau eruption in particular. Dust storms can also produce this phenomenon. Ranchers in the Midwest of the USA have reported St. Elmo's fire on the horns of cattle during dust storms.

There have also been reports of this discharge running along the ground in front of an approaching tornado. As we shall see in a later chapter, peculiar and spectacular luminous phenomena are not at all uncommon in tornadoes and St. Elmo's fire – or something closely akin to it – may explain some of these.

Though probably not St. Elmo's fire in the strict sense, a curious observation from the White Sands National Monument on May 11, 1971 (and reported in the prestigious science journal *Nature* the following year) deserves mentioning here. During a thunderstorm (apparently dry and with much wind blown sand in the air) electric sparks a few yards in length were observed extending straight upward from the tops of several sand dunes. The duration of the "sparks" was not stated in the report, but from the general description it seems that this phenomenon bridges some of the accounts of miniature lightning discussed earlier and the phenomenon of St. Elmo's fire. Clearly, atmospheric electricity can manifest in some strange ways!

The "fire" itself is harmless, but it is apt to accompany phenomena which are anything but harmless – thunderstorms, tornadoes and volcanic eruptions. The "fire's" appearance on an object necessarily means that electrical potentials are high at that object's location and that alone can mark it out as the site of a lightning bolt!

Electric Glows on the Ground and Other Anomalies

From time to time, strange glows are reported that seem to be similar in many respects to classical St. Elmo's fire, yet differing from in by being more widespread in area and frequently of rather short duration. Moreover, these luminous occurrences have been reported to move rather rapidly, sometimes with a "rippling motion" across the ground for significant distances.

Thus, a report from Boscastle, England, dated November 30, 1858, told of a "bright and powerful light" passing the witness "a little quicker than the ordinary pace of a man's walking." The light was seen by others as well, who described it as "coming in from the sea and passing up the valley like a low cloud." The night was one of heavy rain, but no mention is made of lightning.

Also in England some 27 years later, waves of light were seen to sweep across a street, enveloping four people along the way. One of the people involved remarked that his left hand received a slight electric shock at the time. He also estimated the velocity of these light waves as about "30 miles an hour" and mentioned a solitary peal of thunder heard some 3 min later. Another witness, situated outside of the light waves, described the appearance as a "luminous cloud running up the avenue with a wavy motion. When it reached [the group of men] it rose off the ground and passed over the bodies of two of them, casting a sort of flash on their shoulders." This witness estimated that "The distance traversed [by the luminous phenomenon] was about 20 yards, and the time occupied between 2 and 3 s."

The following two observations have each been described as instances St. Elmo's fire, but if that description is accepted, it certainly broadens the scope of electrical manifestations included under this term.

The first, once again, is from England and was published under the simple title "St. Elmo's Fire" in *Nature* 1912.

The account reads,

> Suddenly, without the slightest warning, there appeared an area of faint electric-blue light, almost circular in shape and about 70 yards in diameter, which covered the plateau. The edge of this area was not more than 10 yards from where the observations were made. The whole electric field seemed to be three or four feet above the ground-level and was in a state of intense agitation. Within the general blue ground there appeared flashes of a more decided blue, very similar in character to forked lightning, but not nearly so distinct [does this remind you of the miniature thunderstorm lightning flashes?]. Sounds of two distinct types accompanied the agitation. The first consisted of whistling sounds, like that of numerous long-lashed whips swishing rapidly through the air, or perhaps that

of the whistle of bullets. These sounds seemed to be associated with the general field of fainter blue. The other sounds consisted of the characteristic crackle of electricity, and these became so numerous as they approached a climax that they resembled a magnified rustle. These cracklings seemed to be associated with the forked discharges, and were probably due to the more distinct flashes coming into contact with the bushes which surrounded the plateau. The phenomenon lasted about fifteen to twenty seconds, and disappeared as spontaneously as it had arisen.

The second account, for a change, comes from Yellowstone Park in Wyoming, USA and was published in the journal *Natural History* for June 1950 under the descriptive title "An Electrical 'Bath' in Yellowstone." It happened the previous September.

The witness, W. B. Sanborn, writes that while observing a violent thunderstorm, he noticed a bluish light coming over the low ridge to the west of Swan Lake. He continues,

My first thought was a fire, perhaps caused by lightning. I watched the ridge for a moment and was amazed to see what can best be described as a hazy patch of blue light coming over the ridge and moving down the hill slope toward the flats around the lake. It was then that I observed a very low-grey cloud moving swiftly above the patch of light. The patch moved through the marshy north end of Swan Lake and caused several waterfowl to rise in hurried flight. The patch was but a few yards away, I noted a sudden calm in the air and a marked change in temperature, as well as what I believe was the odor of ozone. It was then that I realized that the display before me was some manner of static electricity, comparable perhaps to St. Elmo's fire and directly controlled by the low cloud moving above. The patch, which was actually a static field, enveloped my immediate area. To describe the weird feeling caused by viewing the progress of this phenomenon is difficult. It kept low to the ground, actually 'flowing around' everything that it came in contact with, coating it with a strange pulsating light. Each twig on the sagebrush was surrounded by a halo of light about two inches in diameter. It covered the automobile and my person but did not cover my skin. There was a marked tingling sensation in my scalp, and brushing my hair with the hand caused a snapping of tiny sparks.

Sprites, Elves and Other Odities

There has long been a suspicion that the lightning flashing within thunderclouds and between clouds and ground does not tell the whole story of atmospheric electrical displays. From time to time there have been reports of glows around the tops of thunderclouds and there exists an account of a spectacular event in New Zealand where the countryside was bathed in light for 15 min by a large luminous globe perched atop a bank of clouds, from whose base ordinary flashes of lightning were frequently seen. And accounts of seemingly ordinary flashes going straight up instead of down, and of rocket lightning taking off from the cloud tops, are too numerous to be ignored, as are reports from aircraft pilots of various electrical manifestations in the upper atmosphere. Nevertheless, meteorologists were hard to convince of the reality of the more unusual reports at least.

Back in the 1920s, Scottish physicist C. T. R. Wilson predicted that electrical breakdown should take place in the upper atmosphere above major thunderstorms, but it seems that his predictions were not taken seriously at the time. Indeed, the first well documented account (in contrast to the numerous anecdotal reports) did not come until July 6, 1989, when scientists from the University of Minnesota observed – and photographed – odd looking transitory glows high in the atmosphere above active thunderstorms.

These discharges looked quite different from lightning. They appeared as pale and extended objects, sometimes looking more like spectral jellyfish than anything else. The preferred umbrella term covering the various forms that these high altitude discharges assume is "transient luminous events" or "TLEs."

A few years after the University of Minnesota results had established that a real phenomenon existed, the name "sprites" was given to the most widely observed variety of TLE. Although this name originated from the mischievous spirit of the air, Puck, in Shakespeare's A Midsummer Night's Dream, the term had also become popular in the 1980s for certain images generated on what was then a popular type of home computer. The similarity in appearance between these computer sprites and the upper atmospheric events played a role in their nomenclature.

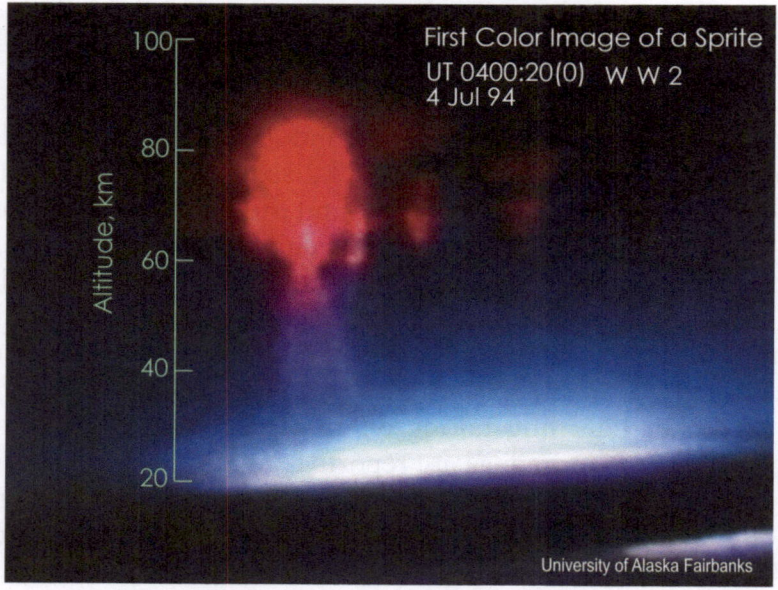

Fig. 4.8. The first color image of a red sprite, obtained during a 1994 NASA/University of Alaska aircraft study of sprites. The event was captured using an intensified color TV camera (Credit: NASA/University of Alaska)

Video images of sprites were obtained from an aircraft study in 1994, after which the term entered into wide circulation.

Sprites typically appear as reddish-orange ("red sprites") or greenish-blue objects from which tendrils hang downward. Arching branches appear above their location and sometimes a reddish halo precedes their appearance. They tend to appear in clusters and occur at altitudes between 50 miles (80 km) and 90 miles (145 km) above the surface of the Earth. Although they have a rather insubstantial look about them, they are nevertheless suspected of having caused some otherwise unexplained high-altitude instrumental balloon accidents.

Sprites appear to be associated with lightning strikes, but another type of high altitude discharge seems not to be directly triggered by lightning and occurs at lower atmospheric levels than the true sprites. These apparitions have been given the more prosaic name of "blue jets." They have been observed extending from the tops of thunderclouds up to the lower levels of the Ionosphere, some 20–30 miles (40–50 km) above ground level. There appears to be a tendency for blue jets to appear above storms in which a significant amount of hail is falling, although so few examples of this phenomena have been observed (and the available sample is heavily weighted by observations of an especially blue-jet-prone storm over Alaska in 1994) that this association remains very tentative. Their color is thought to be due to the blue and near-ultraviolet emission lines in the spectrum of ionized molecular nitrogen.

Ironically, blue jets were discovered from outer space! They appeared on a monochrome video of a thunderstorm in Australia taken by astronauts aboard the *Space Shuttle* on October 21, 1989.

Aerial video monitoring of nocturnal thunderstorms have also found luminous phenomena that may be thought of as blue jets that fail to develop beyond the initial stage. Dubbed "blue starters" for this very reason, these TLEs appear shorter and brighter than fully developed blue jets, and only reach altitudes of about 12.5 miles (20 km) at best.

At the other end of the scale to these diminutive jets, are the rare gigantic jets that can reach atmospheric heights as great as 45 miles (70 km). Only a handful of these have been observed, the first photographed from the Arecibo Observatory on September 14, 2001. This monster jet was estimated to have shot upwards at speeds in excess of 125 million miles (180 million kilometers) per hour! Reaching the ionosphere, it then split in two and moved outward at even greater velocities in a bright burst of light. The few examples of this type of jet that have been observed persist for less than one second and their appearance has been compared to giant trees or carrots.

A different type of electrical discharge has been noted at even higher altitudes over thunderstorms. These have been given the name "elves", an acronym for Emissions of Light and Very Low Frequency Perturbations from Electromagnetic Pulse Sources. The acronym is not, be it noted, a reflection on the size of these phenomena. In truth, they are anything but elf-like in their dimensions. Typical, these high altitude elves appear as dim, flat and expanding glows covering an area of some 250 miles (400 km) in diameter, but lasting for no more than a millisecond. They occur high in the ionosphere, some 60 miles (100 km) above the ground. Like blue jets, they also owe their discovery to the *Space Shuttle*,

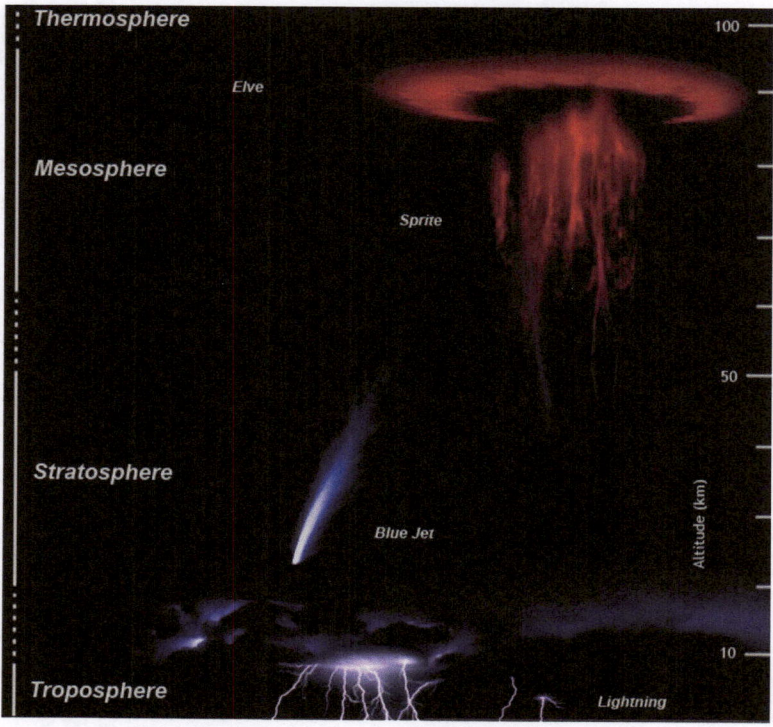

Fig. 4.9. Types of upper atmospheric lightning and electrical discharge phenomena (Credit: Abestrobi 2008)

this time as it orbited high above a thunderstorm off French Guiana on October 7, 1990.

Because of their faintness, the color of elves is difficult to determine, but they are now thought to be reddish due to the excitation of molecular nitrogen through electron collisions. Presumably, the energy driving these collisions is generated by the electromagnetic pulse emanating from lightning discharges in the underlying thunderstorm.

In all of this, one cannot help wondering why this strange upper atmosphere counterpart of lightning was not officially discovered far earlier. Early orbital spaceflights could conceivably have discovered elves – probably the most difficult type of TLE to detect – and sprites are visible at ground level with the naked eye. Lighting watchers or amateur astronomers observing distant thunderstorms under an otherwise clear and dark sky have little trouble seeing and photographing sprites and numerous observations have been made since the early 1990s. How many folk over years or centuries have seen these events and dismissed them as after images or "imaginations"? Surely there must have been many!

Project 9: Sprites

Sprites can be observed with the naked eye from a dark location, but seeing these fleeting electronic ghosts of the sky is not easy. The best conditions for sprite seeking is when there is an active distant thunderstorm, with the top of the thundercloud relatively low over the horizon on an otherwise dark and clear night. Sprites will appear above the cloud tops in a clear sky, but care must be taken that what you see really are sprites and not after-images of the lightning flashes. Happy sprite hunting!

It is indeed strange that in the 1960s astronomical phenomena at the most distant reaches of the universe (quasars, the afterglow of the Big Bang) were discovered by rather sophisticated instruments while events in our own atmosphere and accessible with the unaided eye from a rural backyard remained unknown until 1989! It makes one wonder what other "easily observable" phenomena remain undiscovered.

Sleeks

This thought leads us to a phenomenon that does not officially exist. The only problem, several experienced observers have reported it!

The phenomenon in question concerns relatively faint and swift Most of these reports come from many years ago, but that should not dissuade us from taking them seriously. The reason why *sleeks* (as they have come to be known) are seldom reported these days may simply be that, because they do not "officially" exist, anyone seeing one simply dismisses it as an illusion.

Here is a report by one of the nineteenth century's foremost observational astronomers, W. F. Denning. Whatever we may think of it, this observation should at least be taken seriously simply because it came from such an experienced observer.

Denning writes;

> The meteor was about the third magnitude, but the singularity about it was its marvelous velocity and seeming nearness. It appeared to be in the air, a few yards distant, and I believe its path, extending (as it instantaneously impressed me) over some 16 degrees on the background of the sky, must have been traversed in less than the twentieth part of a second ... Now and then I have observed similar meteors before. They immediately strike one by their close proximity and enormous velocity. They are mere gleams of pale white light, which have little analogy to ordinary shooting stars, and suggest an electric origin, though I do not know whether the marvelous quickness with which they flash upon the eye is not to be held responsible for the sensation of nearness.

Denning goes on to say that they are "somewhat rare", yet he means by this expression only that "one may watch through several whole nights without a single example" – not all *that* rare apparently! He admits to having "witnessed some scores of these meteoritic flashes" during his meteor-watching career.

The above words were written by Denning in 1884. Nearly 30 years later, he again wrote that although "In late years I have not been able to pursue meteoric work to the same extent as formerly" he nevertheless "occasionally caught one of these transient flashes." He described one near Polaris that had a duration of less than a tenth of a second. According to Denning, this one "looked like a mere gleam of light; not a burning missile" similar to an ordinary shooting star.

Other observers have also reported these "sleeks". One report from 1914 involving two of these luminous darts, compared them to the reflections of light moving along lengths of wire. Both sleeks were described as having "terribly rapid" movement.

Sometimes very faint meteors near the limit of naked-eye detect-ability will leave a trail that is more conspicuous than the meteor itself. All experienced meteor watchers will be familiar with these objects, and are unlikely to be tricked. It is difficult to understand how any form of meteor could appear like the moving streaks reported by Denning and others.

Of course, they may really *be* optical illusions, but this is not something to be simply assumed because they do not "fit" with current knowledge. Humanity's store of knowledge only grows by digesting facts which initially appear awkward.

As far as can be ascertained, these streaks of light are not associated with thunderstorms. Obviously, nearby thunderstorms can be eliminated as observers do no watch for meteors in a thunderstorm! But if distant lightning was present at the time sleeks were observed, it is likely that notes to this effect would have been made in the meteor observers' diary. Some mild form of atmospheric electrical discharge cannot be ruled out a priori however. We do not know of any such phenomenon, but if sleeks provide the only evidence for it, that would not be surprising!

Another suggestion poses sleeks as evidence of direct stimulation of the visual channels by cosmic rays. Astronauts traveling to the Moon have reported optical flashes of such a nature, but the Earth's magnetic field shields we surface dwellers from much of this cosmic radiation. Still, cosmic ray particles of high energy do reach the ground and may give rise to something not unlike sleek reports. Possibly the biggest objection to this explanation is that cosmic rays can easily penetrate most buildings and that "sleeks" should be seen indoors as well as outside. But maybe they are. After all, who is going to report a "meteor" in their darkened hallway?

Actually, at least one report does exist! This event, which occurred some time in 1970, is described by the witness as follows;

> I was lying in bed, in a darkened room, when I saw a bright streak of light to my extreme right. Had I seen this streak in the sky I would undoubtedly have called it a meteor ... length 20 degree arc; width ½ degree arc. However, it started from just below the ceiling and disappeared before it reached the floor. It did not move with my eye, but carried on falling as I looked up at it.

One might also wonder if a penetrating cosmic ray can trace an ionization trail through the lower atmosphere and if, at times and under certain circumstances, this may be visible as a glowing "wake" following the particle to ground level.

Be that as it may, meteor observer Vincent Anyzeski noted, from his personal experiences, that sleeks appear more numerous in the early hours of night. At times, he saw up to 20 in an hour, some in his direct line of sight. Their number was highest on three nights of slight auroral activity, which may (or may not!) be significant. In reference to this latter observation, it (at least superficially) may weaken the cosmic ray suggestion, as the flux of cosmic rays reaching Earth is lower during times of high solar activity, thanks to the enhanced intensity of solar wind "blowing" outward in the opposite direction to that of incoming cosmic rays. As increased solar activity is correlated with auroral activity, one might think that sleeks should be more frequent on nights without aurora if they are caused by cosmic rays. An increased incidence during auroral activity might, on the other hand, imply some association with geomagnetic disturbance, although how this could occur is not at all clear.

Anyzeski also noted that these darts of light:

(a) Assumed a variety of shapes
(b) Traveled up to 20° in time intervals as brief as 0.2 s (although some may have lasted up to 0.5 s)
(c) They were from 1 to 10° long and from 0.5 to 3° wide
(d) They flashed into view "in their entirety as to shape"

Project 10: Sleeks

Right here may be an opportunity to prove or disprove the existence of the elusive sleek! Experienced meteor watchers please take note. While monitoring meteor activity, make sure that all participants in the monitoring also note down any sleek-like event, even if they feel that it is only illusory. If significant numbers of potential sleeks are recorded by two or more independent people their objective nature will be hard to deny. Moreover, if their positions are estimated sufficiently accurately for parallax to be measured, their distance from the observers can be derived. In this context, it will be interesting to see whether two observers in close proximity to one another detect a parallax, as that would solve the issue of whether sleeks (if they exist at all) are at the typical distances of meteors or whether they are much closer to the ground.

(e) They "appeared from very few to 20 an hour." (This last does not agree with the supposed rarity of the events. It would appear from this that there were periods when Anyzeski was seeing more sleeks per hour than meteors!).

It would be good if two people had been observing the same patch of sky when a sleek appeared, as the subjective vs. objective nature of the phenomenon would then be distinguishable. Alas, no such instances are recorded, as far as I am aware.

We must leave the issue of sleeks at this point and return to our main theme of lightning and associated phenomena. Let's look now at the puzzling question of the *sound* of lightning; something which form a long while seemed every bit as mysterious and inexplicable as sleeks.

The Sound of Lightning

Everyone knows the sound of lightning. We call it "thunder." But strangely, that is not the only sound that has been reported. We already saw in the previous chapter, how many folk throughout the ages claimed to have "heard" the aurora and how this was long dismissed as illusory. We also mentioned how bright meteors have also been reported to make a swishing or popping sound as they pass across the sky and how this was long dismissed as subconscious association between the meteor and a firework. Presumably, so the psychological explanation assumed, waving curtains of auroral light were associated with waving curtains of fabric, with the mind filling in the missing sounds.

All of these rather pathetic attempts at explaining away well attested observations were finally confined to the trash can by Australian physicist Colin Keay, as we have seen.

Considering the power of a flash of lightning, and its proximity to the observer (a lot closer than auroral curtains and meteors!) it is not too surprising to find that similar sounds have from time to time been reported here as well. Two mechanisms may really be at work in this case. The similarity between many reports of lightning sounds and the auroral and meteoric reports immediately suggests the same or very similar mechanisms, i.e. electrophonic sounds.

Nevertheless, brush discharges from nearby objects may also be a source of some of the reported hisses and crackles associated with lightning and the two may not readily be distinguished.

Project 11: Electrophonic Sounds

Have *you* ever heard lightning "sizzle", meteors "swish" or the auroral curtains rustle? If you have, here are some questions for you.

Did the sound appear to come from the direction of the phenomenon or from your more immediate surroundings?

Was anyone else present with you? If so, did they also hear the sound?

Do you have thick hair, were you wearing spectacles or some other article of dress that might have acted as an "antenna"?

Did you feel any physical sensation (tingling perhaps)?

Was the sound *exactly* simultaneous with the flash or other luminous phenomenon, or did it slightly precede or follow it?

The writer is quite happy to be dogmatic about the existence of this phenomenon. I have experienced it myself and am completely assured that it was real and not some psychological illusion. The lightning flash involved was apparently intra-cloud or inter-cloud (the first more likely I think) as it was very bright, but the thunder was gentle – a "rumble" rather than a "crash" – and not very soon after the flash. I did not see the actual streak of lightning, only the illumination of the night time surroundings. Simultaneously with the flash, there was a relatively loud sound that, as I recall it, sounded somewhere between the static on old AM radio sets and a rush of air being let out of the valve of a hydraulic tire. Perhaps a little nearer to the latter sound.

My experience is reminiscent of an account that appeared in the science journal *Nature* in 1930. The witness writes,

> I did not see the actual flash, only the illumination of the garden through the open window; it was very brilliant and was followed instantly by a noise as though a shower of large water drops had been thrown onto a hot metal plate; this was

followed almost instantly by the thunder. I think there was a slight interval between the swish and the thunder, but it must have been only a fraction of a second. It was also heard by two people in a cottage about 50 yards from the house; one of them likened it to a red-hot poker being plunged into cold water, the other to the sound of the arc when two electric cables are short circuited.

Occasionally, it seems that the sound even precedes the visible flash. A study by S. E. Ashmore, published in *Weather* in 1956, noted that the sound is heard simultaneously with the flash in "almost all" instances, but in rare cases it apparently can precede it. One such example was mentioned in *Nature* (1925). The witness stated that a sound "like canvas being ripped" was heard *just before* the lightning flash. Static on AM radio begins just prior to the visible flash, indicating the generation of electromagnetic radiation at longer wavelengths prior to the emission of light waves, but the sound preceding visible lightning might also be due to intensification of brush discharges similar to St. Elmo's fire just before the principal lightning discharge. Clearly, there is room for much further research here, but at least the psychological explanation has been dispersed like a noxious smog with the inrush of fresh air.

Gamma Rays and Antimatter!

Lightning research in recent decades has uncovered some very energetic accompaniments of lightning. Cartoonists have long shown their fictitious characters displaying their bones in X-ray-like poses when struck by lightning, but it came as a surprise to learn that lightning strokes really *do* emit brief bursts of X-rays! Admittedly not enough to reveal one's skeleton, but the emission of any X-rays at all places lightning amongst the most energetic of natural events. During the 1990s, the *Compton Gamma-Ray (orbiting) Observatory* discovered bursts of the even more energetic gamma rays coming from the direction of the Earth's atmosphere and apparently associated with lightning activity. This link was confirmed by later observations. These *Terrestrial Gamma-ray Flashes* or *TFGs* don't accompany every lightning flash, but it is estimated that about 500 probably occur each day. They are thought to arise from electrons being accelerated to high energies in the strong electric fields of the most powerful lightning strokes. As well as producing gamma-rays, this effect should also produce secondary electrons by interaction with Earth's atmosphere. Electrons produced in this way have now been detected by the *Fermi Gamma-ray Space Telescope* as they follow magnetic lines of force far from their parent lightning strokes.

But *Fermi* appears to have found something even more amazing. In addition to electrons, there is also evidence that a substantial number of positrons (antimatter electrons) are produced as well. When matter and antimatter particles meet, they mutually annihilate and produce gamma rays. The gamma rays produced by the annihilation of electrons and positrons have an energy signature consistent with some of the *Fermi* observations. It remains unknown whether some of these positrons escape into space or whether they add to the Earth's radiation belts. Irrespective of their ultimate fate, it is fascinating to think that a natural process occurring right here on our home planet is mimicking the most energetic events generating antimatter deep within our galaxy!

Thunderbolts

A few words should be said about the term "thunderbolt." For most of us, the word simply refers to the loud crash of thunder accompanying a nearby lightning strike. "Bolt of lightning" is another popularly used phrase for a lightning strike. But in older literature, the "thunderbolt" or "thunderstone" literally takes on a more solid reality.

Thus, according to an entry in *Nature* in 1912, an "aerolite" (an old synonym for "meteorite") was observed to fall at Colney Heath in England during a heavy thunderstorm. According to the witness, the stone's fall was accompanied by an "unusually heavy clap of thunder." The initial description noted that the stone was "fairly deeply pitted, and under magnification exhibits the usual chondritic structure" implying that this was an ordinary meteorite that just happened to fall through an active thunderstorm. Of course, the arrival of a meteorite is always accompanied by sonic booms which sound like "unusually heavy claps of thunder", so it may appear that there is no mystery here – simply an odd coincidence.

However, a follow-up article in the next issue of the journal informs readers that the stone was subsequently submitted to one Dr. George Prior of the British Museum of Natural History who, after examining the specimen, concluded that it was *not* a meteorite!

Earlier issues of the same science journal carried reports of a stone that allegedly fell and buried itself some 8 in. into the gravel during a severe thunderstorm in 1909 and, dating back to 1890, of a "bulky mass" that came roaring through the air "in the midst of a heavy clap [of thunder]" before falling into the ocean about 200 yards from the ship from which this observation was made. According to the witness of this earlier event, the roar of the object's passage through the air was quite dissimilar to the sound of the thunder accompanying the storm then in progress and, upon striking the surface of the water, "spray was thrown fully 40 ft high."

These, and similar, stories have helped perpetuate the legend of the thunderbolt; the stone that is sometimes dropped from a thunderstorm. Various solutions to the mystery have been suggested from time to time. Fusion of dust particles in the path of a lightning flash into a single mass is one idea that has been put forward. Fusion of material on the ground is another, although this could not explain the witness' insistence that the stones truly do fall from the sky. Occasionally, a real meteorite might coincidently fall during a thunderstorm.

But the legend of the thunderbolt extends further back in history than these comparatively recent stories of rocks from the storm clouds. And the ancient thunderbolts seem to have been of a different order to the small rocks mentioned in the abovementioned *Nature* entries. Go back to tales of Thor – the Norse god of thunder – or Jove/Zeus of the Mediterranean lands to gain some idea of how thunderbolts were viewed by the ancients. The thunderbolts that these gods hurled could be pretty destructive. There is a story of Zeus hurling a thunderbolt that wiped out an entire city!

Of course, there is always a danger of reading too much into ancient legend. On the other hand, the astronomical historians Victor Clube and Bill Napier are of the opposite opinion, suggesting that modern folk are apt to read too *little* into old tales. The ancients knew, as well as us, that ordinary lightning is incapable of obliterating cities, so if they had their gods hurling "thunderbolts" capable of that feat, they must have distinguished (according to Clube and Napier) between these "thunderbolts" and ordinary lightning strikes. The "thunderbolts" hurled by Zeus to express his displeasure must have been understood as something a lot more lethal than mere lightning strikes – or even small meteoritic stones that may chance to fall through a thunderstorm.

These authors argue that several thousand years ago, the Earth was periodically subjected to intense meteoric showers unlike anything experienced in recent times. Unlike the sometimes spectacular but completely harmless meteor showers of historic epochs, the hypothetical meteor storms of antiquity were supposed, by these authors, to have had some very large bodies embedded within the streams of small particles. Not only meteorites but *giant* meteorites were a frequent occurrence then. It does not matter that there is no abundance of impact craters several thousand years old. If these large bodies were of cometary origin, they would have been relatively low density objects that burst in mighty atmospheric blasts, not unlike the colossal explosion that flattened large areas of the Tunguska forest in 1908.

The prime suspect, in the opinion of Clube and Napier, is the Taurid meteor complex, a dusty trail associated with the short-period Encke's Comet crossed by our planet in June and November each year. The meteor shower arising from the June crossing – known as the Beta Taurids – occurs in daylight and can only be detected by means of radar, but the November display (also emanating from the constellation of Taurus) is visible at night as a weak but extended drizzle of meteors. The number of meteors per hour is not large (only around 10 even at the shower's height), but what it lacks in quantity, it certainly makes up for in quality. Many of the meteors are bright and some are *very* bright. Relatively large meteoroids inhabit the stream, as fireballs equaling the quarter, or even the full, Moon are sometimes seen with the largest penetrating low enough into the atmosphere to generate sonic booms. Some of the most spectacular Taurid fireballs have the characteristics of meteorites, although no fallen fragments have been found to date. In a typical application of Murphy's Law, the potentially meteoritic ones always seem to fall over the ocean or above extensive wilderness! Hopefully though, one will eventually be recovered; a bona fide piece of Encke's Comet for scientists to study. If I may venture a speculation at this point, I suggest that if or when a Taurid meteorite is recovered, it will probably be an anomalous CI chondrite, somewhat redder than the usual specimens and maybe even more friable.

If Clube and Napier are correct, today's Taurid displays are but pale specters of their former glory. At the time of the early Egyptian kingdoms, not only were the Taurids more prolific in terms of the numbers of meteoroids plunging into our atmosphere, but the number of large bodies – *really* large bodies – was way beyond

Fig. 4.10. Thunderbolt? As the Dutch Meteor Society watched for Leonids in Andalucia, Spain, on 17 November 1995, this Taurid fireball – bright as the quarter Moon – stole the show! Although no meteorites are known to have fallen, the fireball penetrated far enough into the atmosphere to produce audible sonic booms (Credit: Dutch Meteor Society)

anything experienced today. Imagine the impact (no pun intended!) on a civilization if each year in June and again in November fiery bodies poured out of the sky in great torrents, interspersed with dazzling apparitions that were apt to erupt into storms of fire, scorching wind and (we may presume) shining serpents of lightning playing through the clouds of steam and debris raised by the explosions. Let us for a moment picture ourselves on the day when one of these celestial storms was about to break. How apprehensive would we be, not knowing whether this was destined to be the time when one of these super fireballs would explode above our own heads, killing all our family and community; effectively

Fig. 4.11. Comet Encke and its debris trail as imaged at mid-infrared wavelengths by Spitzer Space Telescope 23 June 2004 (Credit: NASA/JPL-Caltech/M. Kelley (University of Minnesota))

bringing to an end not only our own lives but the entire world known to us? Would this not be cause for sacrifice to the gods who controlled these happenings; from whose hands the dreadful "thunderbolts" were hurled?

Studies of the near-Earth space environment over recent decades have discovered an interesting fact. A number of asteroids have been found with orbits surprisingly similar to those of the Taurid meteoroids. Many astronomers believe these to be defunct comets that, together with Encke, were originally incorporated into a truly giant comet deflected into a short-period orbit many thousands of years ago. This "Comet proto-Encke" subsequently broke up into a number (maybe as many as 100) small – though still significant – comets, most of which have now decayed into inert bodies indistinguishable from genuine asteroids. Only Encke retains the capability of visible out-gassing as an active comet.

Against this background, Clube and Napier even go so far as to argue that Encke and its sister comets, which these authors believe to have been bright objects a few thousand years ago, were actually identified with the gods from whom the meteors and "thunderbolts" preceded. Delving into ancient legends, they even find hints in the theogony (birth or generation of the gods) of Mediterranean cultures of observations of the breakup of the original proto-Encke into its next generation siblings.

This is a fascinating subject, but one to which full justice cannot be given in such a short space. Let us just say that, in the present writer's opinion, a lot of what these authors have to say stretches the interpretation of old legends to the limit. A case in point is the suggestion that the pillar of fire and cloud that Moses followed was the tail of a daylight Encke's comet. During the nineteenth century, Encke was consistently seen with the naked eye, although it was never a conspicuous object. But during the twentieth century, naked-eye sightings were few and far between. Only on very good returns was the comet sighted without optical aid, and even then only marginally. In the early 1960s, Professor Fred Whipple plotted the fading intrinsic brightness of the comet and concluded that it was destined to fade out totally by 1999. Looking back the other way, it seemed that Encke should have been brighter than Halley's Comet during the first few centuries of our era. Maybe it *was* a daylight comet in Moses' day!

These predictions were not, however confirmed. On the contrary, the comet survived the 1999 demise date without a problem and at its most recent return in 2010, showed a peak intrinsic brightness not dissimilar to those around the time Whipple made these predictions. Moreover, a detailed search of ancient Chinese records failed to find any comet that could be identified with Encke in 2,000 years of chronicles. Even if Encke had been only 1% as bright as predicted 2,000 years ago, it is hard to see how these ancient astronomers could have consistently missed it. Clearly, the long-term behavior of this comet is more complex than a simple progressive fading model suggests. The rotational axis of its nucleus is thought to wobble quite severely over periods of decades or centuries and this could have a strong effect on its behavior. For example, during the nineteenth century, its axial tilt may have been such that an especially icy region of the nucleus was consistently exposed to full sunlight when the comet was at it closest point to the Sun. If the tilt changed in such a way that this no longer happens, the comet will naturally be less active and therefore fainter than before. The same region of its nucleus might be just as capable of strong activity as it ever was, but if it is not exposed to the full blast of solar heat, that potential will no longer be fulfilled. It is even possible that Encke goes through long periods of dormancy, so maybe the only reason it differs from the other Taurid "asteroids" today is because some event (meteoroid impact?) "woke it up" in recent centuries.

There is some evidence of greater Taurid activity in the past however. An examination of historical fireball records yields evidence of a higher rate of Taurid fireballs in the Middle Ages, but that does not prove the stream was *radically* stronger a few thousand years ago.

It certainly seems that the objects inhabiting the Taurid stream come in a wide variety of sizes. At one end there are the flying mountains of asteroids, while at the other we have meteoroids of gravel size and fireballs up to around basket-ball dimensions. It is very reasonable to think that the intermediate ground is also well populated by objects ranging from automobile size, through house size to football-field size and larger, so the prospect of an occasional giant meteorite arriving with the Taurid stream is not in itself farfetched. But Clube, Napier and some other astronomers go further and suggest that there exists a "core" within the stream where dangerously large objects concentrate and that it was this core that intersected Earth's orbit about the time of early human civilization.

Evidence for this "core" is less than overwhelming. A series of ton-sized meteoroid hits on the Moon in 1975 is often brought forward as the strongest evidence that concentrations of large bodies exists in the stream. However, although the timing of the lunar impacts is consistent with a Taurid association, the orbits of the impacting bodies are unknown and therefore we cannot be sure that they were really Taurids. Still less can we be sure that they formed part of a core of large objects within the stream. They may simply have been fragments of a small disrupted asteroid, which may or may not have been associated with the Taurid complex!

Yet, whatever the truth or otherwise of the entire Clube/ Napier thesis, there must have been times when ancient peoples witnessed devastating meteoric explosions of the Tunguska variety. Maybe some ancient villages or cities really were destroyed by events such as these, leaving a racial memory that was handed down to succeeding generations in what we like to call legends. We may also understand how later generations (who may not have witnessed such large meteoritic events) had these happenings explained to them in terms with which they were familiar, i.e. meteoric explosions looked and sounded something like lightning strikes – only much bigger. In this way, the thunderbolts of Zeus became relatively domesticated as ordinary lightning.

Extraterrestrial Lightning

Lightning is not confined to Planet Earth. In fact, there are lightning storms occurring elsewhere in the Solar System that would put anything that our atmosphere can produce to shame!

Venus, the planet nearest our own, was suspected of producing lightning ever since bursts of radio waves were picked up by early space probes. However, the occurrence of Venusian lightning remained controversial until the European Space Agency's *Venus Express Mission* observed a clear signature of electrical discharges in the form of 100 Hz pulses lasting for just a fraction of a second.

These are almost certainly due to lightning – most probably similar to the cloud-to-cloud flashes on Earth – but actual images of these events continue to prove elusive. There are other mysteries about Venusian lightning as well, which we will take a closer look at in a little while.

First however, let us turn our attention away from the Sun to that giant of the Solar System; Jupiter.

As the owner of even a small telescope well knows, this planetary colossus is one of the real showpieces of the sky. Although orbiting the Sun at a distance of approximately five times that of the Earth, the planet is so huge that it easily appears as a disc in moderately powerful binoculars and even a 2.4-in. (0.6-cm) refracting telescope readily shows the colored bands of its cloudy atmosphere. Sitting amongst these belts is an atmospheric feature of amazing durability, to say nothing of its enormous size. The Great Red Spot, as it is known, could swallow two planets the size of Earth! Winds of terrifying strength (around 400 miles per hour!) whip around its outer rim, yet there appears to be little exchange between its inner region and the surrounding atmosphere. It is now known that the Spot towers above the cloud belts.

Fig. 4.12. Lightning on the night side of Jupiter, photographed by the Galileo spacecraft (Credit: NASA/JPL)

Fig. 4.13. Great red spot of Jupiter, 25 February, 1979 as imaged by Voyager (Credit: NASA)

Though sometimes referred to as a hurricane, the "Spot" is better described as a great anti-cyclonic eddy in the planet's atmosphere. Maybe it's closer terrestrial analogues are oceanic eddies rather than hurricanes. Is it too much to see the "Spot" as a jovian Sargasso Sea?!

As Jupiter has no true surface, lack of friction in this fluid environment means that this particular system has been around for a very long time when compared with the cyclones and whirl-winds of Earth. It was certainly around in the 1800s, becoming unusually conspicuous in 1878, and there is a strong possibility that it is the same feature noted in drawings of the planet as far back as 1631. Over the years, its visibility has waxed and waned. During the early 1960s it was quite prominent in small telescopes and was conspicuously reddish in color. The present writer recalls seeing it plainly in a 2.4-in. telescope at that time. It looked not unlike an eye (slightly bloodshot!) peering out from the jovian clouds. In more recent times it has appeared paler and less obvious and has assumed a more circular, or at least less oval, appearance. Nevertheless, it remains within the range of modest instruments and is not likely to disappear any time soon.

Project 12: Jupiter's Great Red Spot

Sometimes it is obviously red and easily visible even in very small telescopes. At other times, it is paler and a not nearly as obvious.

If the Red Spot is turned toward Earth next time your astronomy club holds an observing night, that would be a good opportunity to see how small a telescope is needed to find the Spot at the present time. A number of observers with telescopes of different sizes might like to try for it. What is the smallest aperture that shows it? If you were naming the feature, would you be happy to call it the Great *Red* Spot?

Jupiter is certainly a turbulent and stormy place. Winds having velocities typical of tornadoes on Earth regularly whip through its atmosphere and powerful updrafts surge from its infernal depths, transporting material high into the outer layers of its atmosphere. Water vapor is carried upwards by these powerful currents and, just as on Earth, it condenses into clouds when it reaches levels of suitably low temperature. This happens at atmospheric pressures some five or ten times greater on Jupiter than on Earth – around five atmospheres as compared with just 0.5–1 on our home planet.

Jupiter was the second planet after Earth where lightning was definitively seen. Being beyond our orbit, Jupiter only reveals its daylight side to us (it is always "full", we might say) and this prevents lightning from being seen through terrestrial telescopes. But once spacecraft ventured beyond its orbit and its night-time side brought into view, suitably Jupiter-sized thunderstorms were not long in being uncovered. Interestingly, these thunderstorms are confined to the upsurging water clouds and as such are not all that dissimilar to the familiar thunderstorms of Earth, in so far as the mechanisms producing the lightning are probably much the same on both planets. The clouds of Jupiter (and also those of Saturn) are mostly *not* composed of water, but of materials such as ammonia and mixtures of ammonia and sulfur. Yet, these are not the jovian clouds that flash with lightning. In this unearthly atmosphere, the Earth-like phenomenon of lightning is confined to the essentially Earth-like watery clouds!

With respect to their size however, jovian and terrestrial thunderstorms are very different. A typical thunderstorm on Jupiter is around 620 miles (1,000 km) wide and lasts for the equivalent of three or four terrestrial days. The towering thunderclouds are around 60 miles (about 100 km) high. In general, lightning flashes are more powerful on Jupiter than their counterparts on Earth, although the very nature of the planet prevents there from being such a thing as cloud-to-ground lightning. There simply is no "ground" in any real sense. But the stronger gravity of Jupiter means that rain falling through the thunderclouds travels much faster than terrestrial rain and the "air" through which it falls is very light. The principal gas in the planet's atmosphere is hydrogen, the lightest of all the elements. By comparison, water is a heavy compound. Fast-falling rain means greater friction and this, as any child who has ever experimented with a plastic ruler and a piece of cloth knows, means a greater electric charge and bigger sparks.

The tops of thunderstorms on any planet are set by the tropopause or boundary between troposphere and stratosphere; the point at which the convective lower air meets the quiet upper layer. This occurs where the pressure falls to the equivalent of one tenth the pressure at Earth's surface. On the other hand, the bottoms of thunderclouds occur where water condenses from vapor to cloud droplets. On Jupiter, this happens at the level where atmospheric pressure is five times that of sea level on Earth. The difference between these levels means that Jupiter's thunderclouds tower much higher than the ones with which we are familiar. A cumulonimbus tower on Jupiter goes up over 60 miles (around 100 km). This is around three times the height of the largest of Earth's towering thunderclouds.

Yet, for all its power, Jupiter's lightning occurs less frequently than its terrestrial counterpart, so the total energy generated by lightning on Jupiter and Earth is roughly the same.

The weather on Jupiter is not, however, equally stormy all the time. Every 15–17 years or thereabouts, there is an outbreak of unusually powerful thunderstorms near latitude +23° on the globe of Jupiter. Some of these have been monitored as moving at forward speeds as high as 370 miles (around 600 km) per hour, indicative of being driven forward by atmospheric winds of this velocity. Despite this latitude being the site of periodic outbreaks of powerful storms, it is not the only region where lightning occurs on Jupiter. Thunderstorms can form just about anywhere on the king of planets.

Beyond Jupiter, the beautiful ringed world Saturn follows its slow course around the distant Sun. Because of its brilliant ring system, this planet has always been a favorite show-piece object for owners of small telescopes who wish to amaze their non-astronomical friends with the beauty of their hobby. The rings (except when viewed edge on) are amazing and unlike most astronomical objects, the planet retains its beauty even when viewed at the ridiculously high magnifications which certain manufacturers and vendors of small telescopes continue to claim for their merchandize. Any experience with telescopes of 2.4 or 3 in. will soon discover that advertising claims that these instruments will take magnifications of 300 diameters and above are (to put it charitably) exaggerated. Jupiter at these powers looks more like a fuzzy comet than a planet, but – amazingly – Saturn continues to be striking. Believe this from a personal experiment racking a 2.5-in. refractor up to 300 power with the help of a Barlow lens!

Yet, take away the rings and Saturn would look pretty bland in small telescopes. No Jupiter-like bands of obvious color gird its globe and prominent long-duration analogues of Jupiter's Great Red Spot are conspicuous only by their absence. Slightly larger telescopes do reveal bands, but they are pale ghosts of their jovian counterparts.

Transitory spots are infrequent and normally quite inconspicuous. The exceptions are the rare great white spots which erupt on the planet about once every 30 Earth years or so, that is to say, once every saturnian year. These have been noted around the time of the summer solstice in the northern saturnian hemisphere in the years 1876, 1933, 1960 and 1990. The first of these became visible in a 2.4-in. refractor and was used to determine the planet's rotation. The Great White Spot of 1933 is the best known of the series (until 1990 at least) and was discovered by William T. Hay, a comic actor who in private life was also an accomplished amateur astronomer. A very youthful Patrick Moore observed it using only a 3-in. refractor, following it for a few weeks before it became invisible. The most recent one in 1990 gradually elongated along the parallel of latitude as it slowly transformed from a white "spot" into a white band, eventually stretching around the entire circumference of the planet before fading away. We can presumably expect another major white spot to form sometime around the year 2020 when summer once more returns to Saturn's northern hemisphere.

Apart from these rare great spots, the comparatively featureless face of this planet was long interpreted as implying that conditions on Saturn are a lot quieter than those on Jupiter. Writing in the early 1960s, Patrick Moore's assessment that "Saturn is a quieter world than its giant brother" pretty much sums up the general consensus of astronomers of that era. But like so many notions of conditions on our neighboring worlds formed prior to the age of space probes, this picture of a relatively peaceful Saturn took a major blow with the arrival of our space craft.

We now know that the apparently calm visage of the planet is due, not to genuine quiescence, but to an obscuring veil of high-altitude ammonia-ice crystal haze. Beneath the haze, lies an atmosphere that turns out to be a roiling, boiling ocean of turbulence, wilder than anything seen on Jupiter. Beneath that benign looking veil of haze, winds roar at an incredible 1,116 miles (1,800 km) per hour!

By the way, the winds of Saturn are not the strongest within the Solar System. Rather surprisingly, Neptune appears to be the windiest member of the Sun's family, with velocities of over 1,200 miles (about 2,000 km) per hour measured in its jet streams. Yet, even this is but a zephyr compared with the gales of extra-solar planet HD 209458b. This is a (broadly speaking) Jupiter-like planet orbiting close to the star HD 209458. Because it transits the parent star as observed from Earth, astronomers have observed absorption lines CO and other species as the light of the star passes through the outer reaches of its atmosphere. From the Doppler shift of the CO line, wind speeds of around 4,400 miles (7,000 km) per hour have been monitored. Theoretical calculations agree that velocities of this order should exist on this planet, which is gravitationally locked such that one hemisphere always faces its sun.

The extreme heat of the sunlit hemisphere, and the extreme cold of the other, creates something like a "sea breeze" on steroids! By comparison with these winds, Earth's strongest recorded natural velocity is the "mere" 318 miles (509 km) per hour that occurred during a tornado in Oklahoma on May 3, 1999.

But back to Saturn! In common with Jupiter, Saturn also has its thunderstorms. By contrast with the storms of Jupiter however, Saturn's thunderstorms appear to be principally confined to latitude –35°, though exactly why this should be the planet's "storm ally" is not at all clear. It may be nothing more than a temporary seasonal effect, with the latitude of greatest storm activity migrating during the saturnian year, but observations have not yet continued long enough to determine whether or not this is so. In any case, some of the storms that brew there are incredible. For instance, the Cassini spacecraft monitored one such storm from November 2007 until July of the following year; a span of 7.5 months. Nothing on Jupiter – much less on Earth – has come even close to this duration. But even that was not the most persistent storm yet monitored. A huge new thunderstorm was noted in the middle of January 2009 and continued raging until October of that year. Lightning flashes were observed both visually and via bursts of radio waves from this storm and their energy was estimated as being some 10,000 greater than corresponding flashes on Earth. The storm cell itself was around 1,850 miles (or 3,000 km) in diameter. This is close to 86% the diameter of our Moon! We may wonder what the thunder would have sounded like, had somebody been there to hear it.

Ten storms were monitored by Cassini between July 2004 and the end of 2009, all of them gargantuan. One peculiar shaped storm found in 2004 has been given the name of the *Dragon Storm* because of its peculiar form. Lightning was not actually seen in that storm, but bursts of radio noise nevertheless betrayed its presence.

Yet, despite the violence and longevity of the planet's storms, lightning on Saturn is still not as frequent as it is on Earth, which appears to be shaping up as the real spark-plug of the Solar System. In general, only one thunderstorm appears to be active at a time on Saturn, a very different situation to that of our home planet where between six and eight *thousand* are sparking away at any one moment.

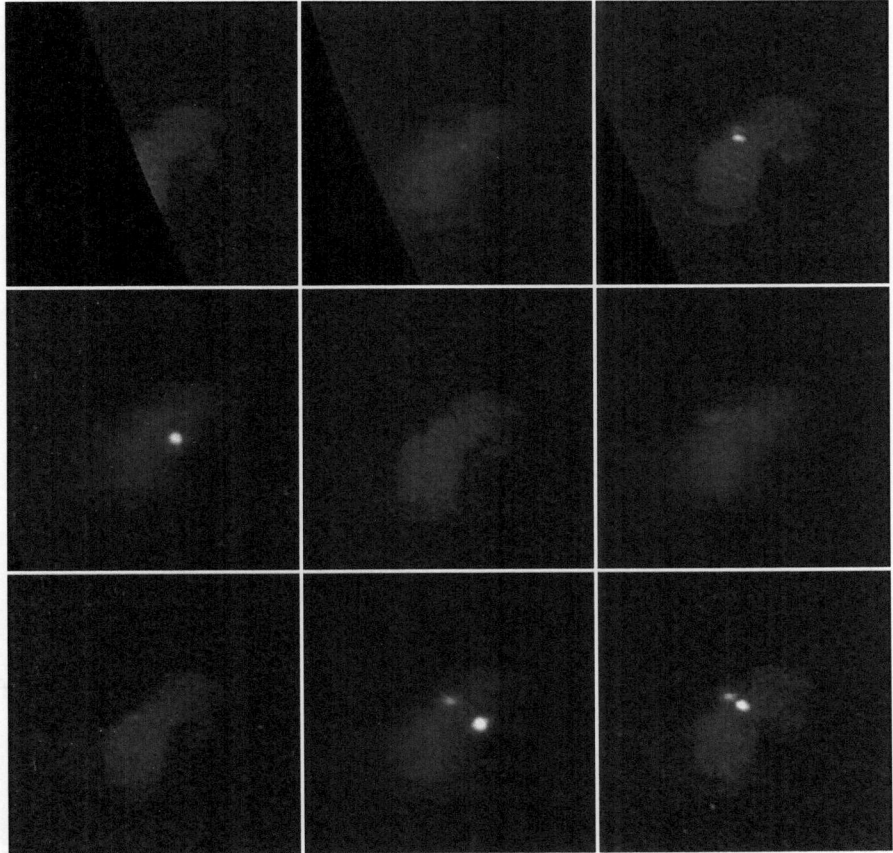

Fig. 4.14. Lightning on Saturn as photographed by Cassini spacecraft (Credit: NASA/JPL-Caltech/SSI)

Cassini data also reveals that thunderstorm clouds on Saturn are of two basic types. They are either quite bright or unusually dark. The luster of the first variety is due to the presence of ammonia ice reflecting back the Sun's rays. The dark ones, on the other hand, are rich in carbon, apparently deposited from carbonaceous compounds over time through the action of the lightning.

Before leaving Saturn, we must say something about its largest moon, suitably named Titan. In many respects, this is the most Earth-like alien world in the Solar System. It has lakes, rivers, rainfall, mountains, sandy deserts and many of the familiar features of home. One of its lakes – Kraken Mare – vies with Earth's

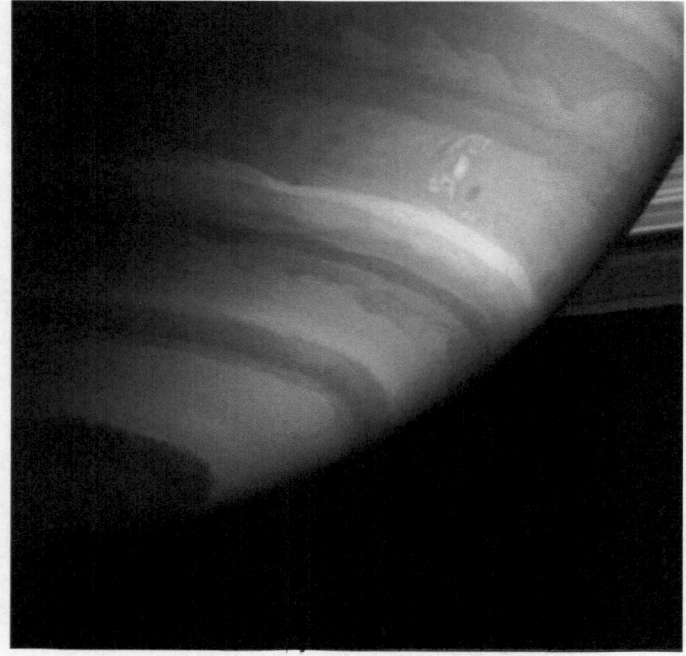

Fig. 4.15. The "Dragon Storm" raging in the southern hemisphere of Saturn (Credit: NASA/JPL/Space Science Institute)

Caspian Sea as the largest in the Solar System. Even its atmosphere (denser than our own, by the way) is comprised largely of nitrogen, unlike that of any other of the Sun's retinue other than Earth. And yet, in other respects Titan is completely alien. Its rivers run, not with water but with liquid methane and other hydrocarbons. Its lakes are similarly composed. Towering storm clouds have been observed in images from the Cassini spacecraft. When these finally dispersed, the land over which they formed was transformed from a dry plain to a lake. Our spacecraft had observed a flood on another world, but once again, the liquid was not water. It was liquid methane! The storm clouds were composed of methane droplets which coalesced together and fell as methane rain, flooding the dry lakebeds on Titan's surface. We can picture this liquid hydrocarbon rain falling in big oily drops which, because Titan's atmosphere is denser than ours and its gravity less, fall more slowly than their aqueous counterparts on Earth.

Fig. 4.16. Saturn's giant moon Titan, 18 October 2010 showing streak of bright clouds near equator, possibly indicating the arrival of spring in the moon's northern hemisphere. The Cassini spacecraft was about 1.5 million miles from Titan when this image was taken. Saturn's rings may be seen faintly at top of image (Credit: NASA/JPL/SSI)

The methane storms of Titan have many features in common with the thunderstorms of Earth. So much indeed, that they are sometimes actually referred to as thunderstorms. The towering convective clouds of these storms are truly methane analogues of the cumulonimbus of Earth and some apparently even reach to altitudes where methane freezes in the moon's atmosphere. They would look very like the thunderclouds with which we are familiar. Yet, to date not a single flicker of lightning has been observed on Titan! Radio bursts recorded early in the Cassini mission and initially thought due to lightning, apparently have another source; most probably related to interaction between the moon and the magnetosphere of Saturn. As far as we can tell, the storms of Titan are "thunderstorms" sans thunder and lightning!

It is just possible that electrical discharges of low intensity and small size could escape notice and the possibility that very occasional flashes of full-sized lightning do happen. Yet, there is no real evidence to support either contention. At least, not at the time these words are being written. Perhaps the situation will have changed by the time you read these lines but at present there is nothing to say that Titan is other than a lightning-less world.

The apparent absence of lightning in Titan's methane storms, appears to be telling us something very important about the genesis of this phenomenon. Lightning may not be possible in clouds of just any composition. Maybe methane does not allow it. Jupiter and Saturn both have water in their atmospheres, whereas Titan does not. Is water necessary for lightning?

But remember Venus? Lightning apparently snakes through its atmosphere, even though the water content is very low there. The droplets comprising the clouds of that planet are sulfuric acid, although there appears to be a component of solid particles whose nature has yet to be properly determined.

Recall also that, even on Earth, lightning has been observed in "dry" circumstances. If the report quoted earlier is correct, occasionally large dust devils would seem capable of generating lightning in their whirling columns of dust and debris. Then there are the volcanic "dirty thunderstorms" and the various electrical discharges (though not necessarily "lightning" in the narrowest sense of that word) which sometimes accompany dust and sand storms. Maybe Venus too is teaching us a little more about the variety of circumstances under which lightning is produced.

With the discovery of extra-solar planets, the potential study of planetary environments has widened far beyond what could have been imagined a very few decades ago. Even now, preliminary studies of the atmospheres of planets orbiting stars light years distant are being made. The first studies of the atmospheres of these far-flung worlds have already been announced and even some of the highly exotic meteorological conditions prevailing on a few of these worlds have been determined. Does lightning happen on some of these strange planets and will it ever be observed, either directly or indirectly? Observation will not be easy, but almost certainly there are extra-solar thunderstorms aplenty, though not on every planet out there just as lightning is not present on all of the worlds of our own system. But what are the conditions under which this lightning occurs? From the results thus far obtained from Solar System studies, it would appear that most of the thunderstorms in the Universe are probably not too unlike those happening right here on Earth. They may be larger, of longer duration and the lighting may be of a different order of magnitude from our own familiar summer fireworks displays. But in all probability, most of the lightning storms accompany clouds of water and water ice, just as they do on Earth and Jupiter. Some extra-solar lightning flashes might rip through clouds of volcanic dust (as some flashes do on Earth) and some may be generated by whatever mechanism works on Venus, but water based thunderstorms are likely very common. Or, expressing it the other way around, where there are planets spawning clouds of water droplets and/or ice, chances are there will be lightning. Thunderstorms rumble and flash throughout the Universe. Now that's an interesting thought to ponder when next you hear the drone of a distant storm!

5. Tornadoes and Other Whirling Winds

Those Fearsome Tornadoes

If the reader will excuse some wanderings down memory lane, let me recall an early spring afternoon when the writer was just a child living with his parents on the Central Coast of New South Wales.

The day had begun unsettled, with distant thunderheads visible from morning and the occasional role of thunder heard at closer range by early afternoon. Still, nothing very dramatic appeared to be in the offing until relatively late in the day when gathering clouds to the west – seemingly illuminated by that ominous greenish glow that traditionally bespeaks the foulest of weather – began their coastward march. As the storm drew closer, one could be excused for thinking that the peals of thunder were showing determined aggression. Each clap began sharp, and before it rolled away, a new and even louder peal replaced it. So the thunder was almost continuous; and seemed conspicuously "angry." This was going to be a bad storm!

Then the rain began. Wind increased and hail started to slam against the wall and windowpane. These were no neat little round hailstones either. They were flat and irregular slabs of ice, the size of the palm of a persons' hand. It was then that something strange happened. The air pressure within the house must have plunged, as the writer clearly recalls his ears feeling full and then "plopping" as they do in a vehicle ascending a high hill. Doors within the house swung open without any perceptible draught and kitchen cupboards popped open. Then, all became quiet. Eerily quiet.

The quiet was soon broken however by a neighbor and close relative banging on the back door to see if we were all right. On emerging, we saw what he meant. Not far from the house, a portion

D.A.J. Seargent, *Weird Weather: Tales of Astronomical and Atmospheric Anomalies*, Astronomers' Universe, DOI 10.1007/978-1-4614-3070-4_5,
© Springer Science+Business Media New York 2012

of a nearby school building lay astride the road. A little further behind us, a small house (thankfully, unoccupied at the time) had effectively disappeared. All that remained were the floorboards and walls stacked one upon the other *beneath* (not on top of!) them. Further still down the town, the scene was more reminiscent of a bombing raid than the aftermath of a storm. Houses and shops were in ruin. Electricity wires looked like balls of string that had survived the attention of a very playful cat!

To the west, beyond the furthest shore of the large coastal lake whose opening to the ocean gives the writer's home town of *The Entrance* its name, an area that in those days was mostly forest land with the occasional farm was spared no mercy by the storm. Two cows were reportedly lifted over 100 ft into the air like kites, before fatally plunging to the ground. An area of virgin forest was cut down as if by a gigantic scythe. Grass was pulled out of the ground, trees stripped of branches and bark and hurled like dressed poles for up to a mile through the air. One witness to the devastation found a boulder, at least a yard across, pulled from the earth and lying beside a crater where it had once been securely fixed into the ground.

A resident near the affected area heard the sound of what he thought was a squadron of jet aircraft flying overhead. Looking outside, he saw, not a flight of aircraft, but an enormous whirlwind approaching his home. He recalled thinking "My son is having his seventh birthday tomorrow, and I won't be there. I'll be dead!" But the whirlwind turned along a nearby ridge and missed his house. His son turned seven the next day. Happily, his father was still to celebrate with him!

Another resident was also alerted by a sound like screaming jets and, looking outside, was amazed to see the sky apparently filled with parachutes! "What are these idiots doing parachuting in weather like this?" he thought (or words to that effect!). Then he looked more closely. The "parachutes" were in reality trees and parts of trees falling from the clouds!

The area had experienced that most terrifying of meteorological phenomena; a tornado. From what was learned later, it became apparent that, despite the devastation that it caused, the storm as we experienced it was but a shadow of it's former self. Had it hit the town with its full fury, it is unlikely that you would be reading this book!

NSSL Photo

Fig. 5.1. Tornado at Alfalfa, Okalahoma, 22 May 1981 (Credit: NSSL)

As defined by the official British Meteorological Glossary, a tornado is a whirlwind of relatively small diameter that is often associated with a more or less violent thunderstorm. American meteorologists add that tornadoes are associated with cumulonimbus (thunderstorm) clouds.

This is broadly true, although not every whirlwind associated with a thunderstorm is a tornado (thunderstorms sometimes generate whirlwinds of a different variety along their gust fronts; but more of this later) and not every tornado is associated with a thunderstorm or cumulonimbus cloud! That is, unless one simply defines a tornado as a whirlwind associated with such a cloud. In that case, phenomena looking exactly like tornadoes, albeit associated with non-cumulonimbus types of cloud, must be given some other title! This too, we shall take up again in a little while.

Unlike the small fair-weather whirlwinds that skip and play over ploughed fields on a summer's day, a tornado characteristically *descends* from a cloud. This does not mean that the air inside the whirl is descending. Clearly it is not. But the vortex itself is. This apparent paradox is resolved if we picture the tornado as a section of a rising column of air that is being dramatically concentrated and "spun up" into a veritable pipe of wind. The apparent descent of the tornadic vortex simply means that the rising air column begins to constrict at the level of the cloud base and that this constricted section extends downward as more and more of the column becomes included within it. As the column constricts, the circulatory wind increases in velocity and the atmospheric pressure falls dramatically within the centre of the developing – and lengthening – vortex. As pressure drops, the air cools. Artificial applications of this phenomenon range as wide as refrigeration units and Wilson Cloud Chambers, but in the tornado, the process gives rise to an effect that is often seen as being characteristic of these storms; the *tuba*, also known as the *condensation funnel* or, more simply, *funnel cloud*. Actually, a funnel cloud doth not a tornado make. In conditions of low humidity – especially if the tornado is weak – a funnel cloud frequently does not form, yet the storm is still a tornado. Nor is it true to equate the extent of any visible funnel with that if the tornado itself. In short, just because a funnel cloud may not reach all the way from cloud to ground, this does not necessarily mean that the tornado itself is not touching down. To imagine otherwise might be a very serious error of judgment! Usually, a whirl of dust and debris below a pendant funnel cloud will indicate the tornado's presence on the ground, but even the absence of any visible whirl might be deceptive. As a rough rule of thumb (please note the "rough") it is often said that if the funnel cloud extends about two thirds of the distance from cloud base to ground, the tornado itself has already touched down. In dry air however, the funnel cloud may be much shorter than this, if (as already mentioned) one forms at all. By the way, the air pressure at the bottom of the funnel cloud matches that at the cloud base, so in a sense it is a tubular downward extension of the cloud base itself.

To be classed as a tornado, the vortex itself (with or without a condensation funnel) must reach the ground in the sense of maintaining a whirl at ground level strong enough to cause surface damage. "Damage" in this context is defined very broadly and liberally. If your hat blows off while a funnel cloud is overhead, you have been caught in a tornado! "Damage" in the minimalist sense used here simply means that the circular wind at surface

Fig. 5.2. One of several tornadoes that hit central Oklahoma on 3 May 1999 (Credit: U.S. National Oceanic and Atmospheric Administration)

level is sufficient to move leaves, papers, blow dust or in some way disrupt the existing status quo. The very weakest tornadoes do no more than this. The most violent cause total destruction. Most reported twisters lie somewhere in between.

If a funnel cloud is seen aloft, but there is no disturbance sufficient to cause damage (in the very broad sense being used here), the phenomenon is simply noted as a funnel cloud, not a tornado. Of course, the cloud could not exist without the presence of a tornadic-type vortex, but to make it to the records as a full tornado, the vortex must touch down at least briefly.

Many tornadoes do in fact assume the appearance of a funnel cloud extending only part way down from the cloud base, albeit coupled with a column of dust or other debris reaching upward from the surface toward it. Sometimes the two will meet, sometimes not. The first appearance of this tubular vortex of dust (known as the *annulus*) beneath a funnel cloud means that the vortex has reached the surface and qualifies as a tornado. Quite often, the rising dust column encases the condensation funnel like a sleeve, so that it is not always easy to see where one stops and the other starts.

Fig. 5.3. Large funnel cloud, May 24, 2008 (Credit: Christopher Williams)

Fig. 5.4. Tornado at Erick, Okalahoma, 25 April, 1989. No condensation funnel is visible and the tornado's presence is marked only by the column of dust and debris kicked up from the ground (Credit: NSSL)

If a funnel cloud does not form, the disturbance at ground level is the only evidence that a tornado has touched down.

If the tornado forms over a body of water, the annulus is comprised of spray instead of dust and is then known as a *cascade*. The storm also acquires a new name; a *waterspout*. Waterspouts are not, in spite of their name, spouts of water but simply tornadoes at sea or over large inland lakes. Because of the consequent abundance of water vapor available to them, impressive funnel clouds are readily formed even when the tornado itself is very weak. Also, the lack of surface friction over water, relative to that experienced by a land-based vortex, probably aids in the persistent of weak whirls that would quickly be disrupted over land. Even very imposing waterspouts frequently collapse once they reach the shallows. (Meteorological purists will no doubt be cringing at this simple equating of waterspouts with tornadoes over water. In defense, let me say that I realize that I am oversimplifying matters here, but I also reserve the right to speak about waterspout vortices as genuine types of tornado vortices – something with which not everyone will agree. But more about this in a little while!)

The "classical" tornado is born within a supercell thunderstorm, the most dangerous class of storm (if we include within the class the very rare mega-supercell storms) and the ones responsible for the worst of thunderstorm-associated severe weather such as large hail, damaging straight line winds, flash flooding and the like. When conditions are right, they add tornadoes to their list of weapons.

In common with their (generally) less violent brethren (single cell and multi-cell storms) supercells pull in warm and moist air from the surrounding regions and propel it aloft through powerful updrafts. At high altitude, moisture condenses into cloud and rain. As the latter develops, surrounding air is cooled and begins to sink toward the surface as a downdraught. In ordinary non-supercell storms, this downdraught soon overwhelms the updraft and the storm cell collapses. If the storm consists of just one cell (i.e. if it is of the type not surprisingly known as single-cell storms) that is the end of it. If it is of the multi-cell type, new cells constantly form as old ones die away and the overall storm continues to regenerate itself. Either way, the individual cells contain within themselves a sort of self-destruct mechanism which insures that once they reach maturity (i.e. when the rain starts falling) it is all down hill from there.

Not so for supercells! These storms are so large that there is plenty of room for both the updraft and downdraught to operate together, without the latter canceling out the former. On the contrary, each feeds the other, keeping the mighty storm active for hours as it tracks across many miles of countryside.

As surrounding air flows into these huge thunderstorms, it begins to rotate. This initial circular motion may come from a number of sources, the Coriolis force resulting from Earth's rotation plays its role, but more the more immediate factor is the presence of wind sheer. Strong winds blowing over weaker ones can cause the air in between to rotate around a horizontal axis, rather like a pencil rolling along a tabletop. If this spinning tube of air is ingested by the updraft of the storm, it tilts toward a vertical position and sets the updraft rotating. The presence of a jet stream – a relatively narrow but very strong ribbon of westerly wind blowing at an altitude of about 6–8 miles (9–13 km) – strengthens this circulation and updraft. The effect of a jet stream blowing over the top of an updraft is somewhat like a musician blowing across the top of a set of pan pipes, except that the result is not exactly a musical note! The region of rotating air within the thundercloud stretches upward and downward and, in so doing, intensifies even further, in the process pulling in yet more surrounding air to feed the growing vortex. This phenomenon is called a *mesocyclone* and, in its early stages, is typically in the range of 2–10 miles (3–16 km) across. As it stretches and intensifies, it draws downward to less than a mile from the ground (where it is marked by a dark, rotating, lowering of the cloud base known as a *wall cloud*) and up to around 10 miles into the atmosphere, overshooting the thundercloud's anvil and creating a emergent dome typical of really severe storms.

A dangerous feature of a mesocyclone is the presence of active and persistent updrafts and downdraughts in close proximity to each other. This, as we said earlier, is something that sets supercell thunderstorms apart from the other varieties. A very dangerous situation develops if a downdraught feeds into the rotating updraft from the back of the storm. Such downdraughts are known as *rear-flank downdraughts* and as the descending air of the downdraught gets drawn into the ascending warm air of the updraft in the lower portion of the thundercloud, the rotating updraft itself is pulled downward and spun up through stretching. This gives rise to a smaller and more intense region of enhanced convergence and vorticity within the larger circulation of the mesocyclone. If this vortex column stretches downward all the way to ground level, a tornado is born. Tornadoes that form this way – let's call them "mesocyclonioc tornadoes" – include the most violent examples of these types of storm, although, thankfully, not all mesocyclonic tornadoes are of equal severity.

Tornadoes of this variety – and they are the "classical" ones about which most research has been done – are typically seen as dark funnels hanging down from the cloud-base lowerings known as wall clouds. They occur at the back Equator-ward quadrant of a supercell storm, i.e. south-west of the storm center for the northern hemisphere and north west of center for the southern hemisphere, assuming a usual east-west progression of the storm itself.

Like hurricanes, tornadoes also have "eyes" or centers of very low pressure and relative calm, although it goes without saying that the scale is altogether different from that of a hurricane. When a large tornado hit suburban Brisbane in Queensland, Australia, on November 4, 1973, one witness described how a tremendous wind first struck one side of the house, followed by a very brief period of calm which soon gave way to an equally damaging squall of wind striking the opposite wall of the house. In a large tornado, cool air is drawn down this calm central tube by the extremely low atmospheric pressure at ground level, there to turn outward into the primary inflow being drawn in at ground level from outside the vortex. This combined flow is again caught up by the updraft, further enhancing the velocity of the circular winds.

The way is then prepared for a striking phenomenon to take place. Both theoretical and experimental investigation show that if the ratio of the tangential wind velocity to the vertical velocity (known as the *swirl ratio*) attains a certain critical value, the vortex breaks up into a series of smaller vortices skirting around the periphery of the large one like carriages on a carousel. As this ratio increases in direct proportion to the inflow rate, the abovementioned mechanism helps to push the tornado toward developing a multiple vortex structure.

A multi-vortex tornado typically assumes the form of a wide and rather diffuse funnel in which smaller vigorous swirling columns dance around the circumference. These *suction vortices*, as they are known, might be small, but they can pack a terrible punch! As the tornado concentrates the power of a mesocyclone, these mini-vortices concentrate the power of the tornado. The strongest natural winds generated by any meteorological source on this planet are found near the base of the most vigorous of these suction vortices. The most severe tornadoes usually assume a multi-vortex aspect for at least some of their career and many of the oddities noted in tornado damage patterns can be explained by the formation and decay of these tiny but often incredibly intense secondary whirlwinds.

Non-mesocyclonic Tornadoes and Other Whirlwinds

Tornadoes can also form when a weak circulation of air is swallowed by an updraft of a developing thunderstorm (not necessarily a supercell in this instance) or even a rapidly growing cumulous cloud. What began as an innocent circulation of air, possibly originating as a horizontal eddy created by wind shear, once ingested by the updraft becomes intensified dramatically through vertical tilting, stretching and narrowing. The physics at work here is the same principle of *Conservation of Angular Momentum* that causes a ballerina to spin up if she tucks her arms to her side. It is also the same principle that spins up contracting clouds of cosmic dust and gas into protostars and eventually into rotating stars and pre-planetary discs. But in a thunderstorm the effect can be frightening. Imagine a 5-mile-per hour breeze circulating around an eddy 4 miles across. Now think what happens if this is taken up into the updraft of a thunderstorm and the diameter of the circulation reduced to a typical tornadic width of 400 yards. By the Law of Conservation of Angular Momentum, the gentle zephyr is increased to a roaring 160-mile-per-hour (256 km-per-hour) wind!

Tornadoes formed by this latter process are generally described as rather weak, although "weak" when applied to tornadoes must be interpreted rather liberally. Even a relatively "weak" tornado can still be a frightening experience to live through and may result in a good deal of damage and injury. No tornado is to be treated lightly!

Landspouts – (A Rose by Any Other Name Is Just as Thorny!)

As mentioned earlier, some meteorologists are not happy about calling anything other than a mesocyclonic tornado a true *tornado*. Of course, if the term "tornado" is simply defined as referring to a violent whirlwind associated with the mature stages of a supercell thunderstorm, then any other type is ruled out immediately. But surely such a definition is too restrictive, and it will be assumed here that the term is legitimately applicable to any narrow, vertical whirling column of air associated with a cloud or storm.

Fig. 5.5. Landspout near North Platte, Nebraska, 22 May 2004. Note the very small condensation funnel at base of cloud and the large dust tube extending from the ground (Credit: U.S. National Oceanic and Atmospheric Administration)

The type of tornado occurring in developing thunderstorms (i.e. in cumulonimbus clouds that are still growing and have not yet reached the mature stage when rain begins to fall) or in large growing cumulous clouds are known as *landspouts, dustspouts* or *dust-tube tornadoes*. The term "landspout" is sometimes said to have been coined by meteorologist Howard Bluestein in 1985, but it also appears to have been independently invented a little later (in 1988) by R. Brady and E. Szoke in a paper dealing with this type of tornado as found in northeast Colorado. And, long before either application of the term, way back in 1927 H. Busk described a tornado seen in Persia on December 27, 1915 as a "landspout" to emphasize its strong similarity with a waterspout. To make things even more confusing, British meteorologist G. Meaden employed the term for any tornado (even mesocyclonic ones) occurring over land, that is to say, any tornadic vortex other than a waterspout!

Given sufficient humidity, a landspout develops a tube-like funnel cloud that may or may not (and frequently it does not) reach the ground. It is usually relatively narrow and has a smooth appearance, very much like a waterspout over land, as the name indeed suggests. When the funnel cloud does not reach the ground, the presence of the surface vortex is usually marked by an annulus of dust and debris or, in stronger examples, by a tube of dust that may reach the cloud base. In contrast to mesocyclonic tornadoes, the funnel cloud protrudes directly from the cloud base, normally unaccompanied by an ominous wall cloud. At most, the top of the funnel might be surrounded by a "collar cloud" of much smaller extent than a true wall cloud. Where the humidity is very low or the vortex is weak, a condensation funnel may not form and the only evidence of the tornado is a swirling column of dust beneath a developing cumuliform cloud.

Occasionally there is not even that. One such "invisible" landspout was probably responsible for the following embarrassing incident related to the writer by a well-known Australian meteorologist. The story involved a South Australian man who decided to stop at the municipal swimming pool to cool off on a hot summer's day. He undressed to his swim wear, leaving his trousers over a seat beside the pool and dived in. Whilst in the water, the pool and surrounding park was hit by what was described as "a tremendous gust of wind." The lids of rubbish bins popped off, leaves and debris blew hither and thither and the man's trousers became airborne. Red faced (or so we might assume) the man confessed that the last view of his trousers was as they disappeared into a large cumulous cloud building up overhead!

Landspouts are generally smaller, weaker and shorter-lived than their mesocyclonic relatives. Most (though certainly not all) last for less than 15 min or thereabouts and register scores of 0 or 1 on the Enhanced Fujita Scale (EF) of tornadic damage potential (see Appendix). In terms of wind speed, this translates to somewhere between 40 and 112 miles (64–180 km) per hour. Note that the middle-to-upper range of these wind speeds are quite considerable. On the Beaufort Scale, a wind of 73.4 miles (117.5 km) per hour is classified as hurricane force, yet many of these "weak" tornadoes well and truly exceed this level. Moreover, landspouts packing wind velocities within the EF3 range (159–207 miles or 254–332 km/h) have been noted and there is at least one landspout rated as F4 on the earlier Fujita Scale, i.e. packing wind speeds of between 208 and 262 miles (333–419 km) per hour. That is hardly *weak* on any assessment! The landspout clearly has the potential of being a dangerous event and should always be treated accordingly.

Waterspouts

Earlier, I said that meteorological purists probably cringed at my oversimplified characterization of waterspouts as tornadoes over water. I will explain (and hopefully defend) myself in a moment, but first of all, let's look a little more closely at this wonder of wind, cloud and spray.

First of all, as also mentioned earlier, a waterspout is not a column of water, so technically the name is something of a misnomer. A typical waterspout is first noticed, not on the water at all but near the base of a cloud above it, in the form of a short projection that can have a remarkably smooth and even "solid" appearance. Although it might well pass unnoticed, about the same time a circular patch of slightly disturbed water appears on the surface beneath this incipient funnel. The developing condensation funnel, appearing like a translucent length of tube in which the central

Fig. 5.6. Three waterspouts near the beach at Kijkduin, near The Hague, The Netherlands on 27 August 2006 (Credit: Wikipedia)

regions are noticeably free of cloud, snakes downward from the initial projection and, as it approaches the water surface, the circular patch becomes increasingly disturbed until a spray ring forms around its circumference, stretching upward toward the lower end of the funnel cloud. Depending upon the strength of the winds within the vortex, this cascade of spray will either remain a fairly innocuous whirl near the water's surface, reaching nowhere near the height of the funnel cloud, or become a towering funnel partially enveloping the condensation funnel and even reaching all the way up to cloud base. Once formed, the waterspout will usually persist for between 10 and 20 min as it moves slowly across the water's surface. In most instances, small waterspouts of the type just described tend to dissipate if they approach the shoreline.

As you might have noticed, the description given here sounds a lot like that of a landspout; except that it forms over water. The general appearance and the dynamics are very similar and – irrespective of who was first to use the term "landspout" – the reason for coining the word is obvious enough. These tornadoes so closely resemble waterspouts.

Now traditionally, waterspouts of this type (the "classical" waterspout as they might be called) have become known as *fair-weather waterspouts*. This is to distinguish them from the other variety, tornadic waterspouts, which are simply mesocyclonic tornadoes that either formed over water or which began over land and then crossed the coast.

Project 13: Seeking Waterspouts

This is one for readers who live on the coast or shoreline of extensive inland bodies of water.

Waterspouts of the "fair weather" type typically form beneath large cumulus clouds growing over bodies of water that are warmer than the overlying air. Look for any sign of a funnel cloud projecting downward from the flat bottom of a large cumulus. If the funnel is very short, there may not be a cascade of spray on the water's surface beneath, but if it extends downward for approximately one third of the distance, there will almost certainly be a visible vortex at the surface. How long do the waterspouts last? Do they collapse if they come too close to land?

This distinction is confusing. After all, we have just been saying that a landspout is a bona fide type of tornado, so why should the concession not be extended to its marine counterpart as well? The fact that waterspouts appear to form more readily than landspouts and that many fail to make landfall not withstanding. If a landspout is classified as a tornado, what does it become if it moves over water? Conversely, how is a waterspout to be classified if it moves onto land? Especially if it causes significant damage? Stronger waterspouts, often accompanying (non-supercell) thunderstorms do sometimes make landfall and even strengthen as they pass inland. These are always reported as tornadoes, even though they may not be mesocyclonic.

Moreover, the description "fair weather" requires a goodly dose of poetic license. The odd waterspout *has* been seen in fair-weather conditions, even descending from stratiform clouds (!)

but most are associated with fast growing cumulus congestus that develop showers or thunderstorms at maturity, or in rising air before advancing rain squalls. Except for the waterspout itself, these situations are hardly violent weather events, but they are not strictly "fair" either. Perhaps it is less confusing simply to speak of mesocyclonic and non-mesocyclonic waterspouts or (for brevity's sake) simply call any that are not mesocyclonic, "waterspouts."

Waterspouts, as just defined, form most frequently in near-coastal waters. They are not as common over the remote oceans far from land. Although most frequent over tropical and subtropical waters, they also occur quite often in temperate latitudes and are not infrequently seen in the English Channel and off the European coast. Warm currents such as the Gulf Stream flowing into colder regions probably aid their development at higher latitudes. They seldom last longer than about 20 min and most seem to register only 0 on the Enhanced Fujita Scale. In other words, the velocity of their winds is usually of the order of 40–72.5 miles (64–116 km) per hour.

Cold-Air Funnels

This term was coined by American meteorologists J. Cooley and M. Soderberg in 1973, to describe a type of small funnel cloud that forms in cool air well behind a frontal or trough system and which is typically associated with either cumulus congestus cloud in the absence of thundery conditions or non-severe cumulonimbus clouds forming within the circulation of a cold core low pressure system. They also form from time to time in clouds that are best described as fair-weather cumulus and have been noted as rope-like loops of cloud unattached from nearby small cumulus on cool sunny days. Some have even been noted in stratus cloud, the most stable of cloud types!

The conditions under which these funnels form are, at worst, describable as "drizzly with showers" and, at best, clear and sunny with scattered cumulus. Neither circumstance could be described as conducive to severe weather.

Cooley and Soderberg think that these funnel clouds and their accompanying vortices owe their existence to a layer of humid air

Fig. 5.7. High based funnel cloud extending from decaying fair weather cumulus over Northern Texas, May 29, 1994 (Credit: VORTEX Project)

lying beneath one of dryer air. Because water vapor is a light gas, moist air is lighter than dry and a certain instability is generated as the humid air rises through the drier layer. If this is the mechanism responsible for these funnels, Cooley and Soderberg suggest that it could explain why so few reach the ground and why they seem to occur more often in coastal regions where moisture levels tend to be higher at low altitude. Lacking thermal instability at low levels, the air near the ground will normally be too stable to bring the vortex down to surface level.

Cooley and Soderberg do, however, recognize a class of cold air waterspouts which are really cold air funnels that do manage to touch down onto lakes and ocean surfaces. Cold air landspouts – i.e. cold air funnels that *do* manage to reach the land surface – have also been suspected, and undoubtedly do happen from time to time.

Cousins of the Tornado

The whirlwinds about which we have been speaking; mesocyclonic tornadoes, landspouts, waterspouts and even the occasional cold-air funnel that manages to touch down, are all classifiable as "tornadoes" in the sense of being vertical tubular vortices which intensify downward from a cloud base. But not every whirlwind is of this type. The tornado has a number of more or less distant "cousins." Fortunately, none is as destructive as its notorious relative but each is nevertheless an interesting phenomenon in its own right.

Wind Devils

This category includes all whirlwinds that form as swirling updrafts in fair weather conditions. Warmer air near the surface rises through an overlaying layer of cooler air and is set rotating,

Fig. 5.8. Dust devil in Ramadi, Iraq, July 2007 (Credit: Andrew Patton)

probably by the prevailing wind. As the air rises, a column of air stretches vertically and the vortex intensified. When fully formed, the vortex drags in surrounding warm surface air and carries it aloft in a rapidly rotating cylinder. As the air rises, it cools and becomes denser, eventually sinking back toward the surface.

Depending upon the material lofted into the vortex, wind devils can be divided into dust devils (the wind devil *per excellence*, sometimes also called the "dirt devil"), snow devil or water devil. Forest fires generate columns of wind and flame called, unsurprisingly, fire devils. Incidentally, large fires can trigger true tornadoes of the kind known in Japan as dragon twists, but these are of a far more violent nature than the more usual fire whirlwind and are accompanied by the development of fire-induced cumulus clouds or *pyrocumulus*, which in the most extreme instances can develop into fully fledged thunderstorms. On occasions, the rain from these storms even quenches the fire that gave them birth! But that is another story!

Wind devils normally do little more than blow loose debris in circles and raise columns of dust, but larger and more vigorous ones can attain the strength of EF0 tornadoes capable of inflicting minor damage to property. Wind strengths of around 75 miles (120 km) per hour have been estimated in some large ones, but in most instances winds are unlikely to become stronger than about 45 miles (70 km) per hour.

The chief difference between a tornado and a wind devil is the latter's development from ground up rather than descending from a cloud base. But can a *funnel cloud* form in a wind devil?

Large wind devils – the true dust devils – that grow to the violence of EF0 tornadoes could probably form condensations funnels, except that they typically form in very dry conditions where even true tornadoes would be hard pressed to form funnel clouds. Wind devils forming in more humid air – like the little whirls that skip over urban car parks and the like – are simply too weak and do not extend to sufficient altitude to form condensation funnels.

Project 14: Wind Devils

The direction of circulation (cyclonic vs. anti-cyclonic) of wind devils is likely to be influenced by local conditions rather than Coriolis force. But do they show a preference for one direction of rotation? The writer is aware of only one informal study (in the southern hemisphere) which suggested a preference for anti-clockwise or anti-cyclonic rotation, but the sample was probably not very large. If the reader lives in an area prone to small-scale whirlwinds, he/she may like to monitor as many as possible to see if they really do show a preference. If you have a Facebook™ friend in the opposite hemisphere who is also interested in doing some whirlwind monitoring, it would be interesting to compare results. Do small whirlwinds show any preference in direction of rotation and, if they do, does this preference change according to hemispheres?

Coriolis force is more pronounced at higher latitudes. It would be interesting if this project could be repeated by several people at different latitudes both north and south of the equator and their results compared. Is there a tendency for whirlwinds at higher latitudes to show preferential rotation more than those occurring closer to the equator? (see also Project 15).

One more question to ponder. Can a wind devil cross the equator?

If you have a hand-held anemometer and are not adverse to jumping into wind devils, try determining the wind strength in whirls of various sizes. You may also like to try measuring the pressure drop by using a small barometer. Best of luck!

Water devils or whirlwinds occurring over the ocean or inland lakes are more promising and there have indeed been reports of clear-air waterspouts; funnel clouds forming over whirling columns of spray and extending upward into a cloudless sky. The writer even saw one himself at one time. A dry, cool wind change swept up the coast and for a short while a tube-like funnel cloud swept along at the point of convergence of the winds. The funnel was the only cloud in the sky and remained over the ocean a short distance from shore. Although the surface of the water was not visible from my location, a spray cascade would almost certainly have been present.

Gustnadoes

Sometimes short-lived vortices form at the advancing gust front (resulting from the outflow of sinking air) of a thunderstorm or along the leading edge of a squall line. Popularly known as *spin-up tornadoes, gust front tornadoes* or simply *gustnadoes*, these whirlwinds have more in common with wind devils than with true tornadoes. In common with the dust devil, they extend upward from the ground rather than intensifying downward as in a true tornado. On the other hand, some gustnadoes can reach to cloud level and show up as protrusions in the cloud base. The term "spin-up tornadoes", although sometimes used of any gustnado, is more properly used of these rare types.

Fig. 5.9. Gustnado in southeastern Wisconsin, 4 October 2002 (Credit: U.S. National Oceanic and Atmospheric Administration)

Condensation funnels are not normally associated with gustnadoes, although the "clear-sky waterspout" mentioned at the end of the previous section formed at the leading edge of a dry cold front and could probably be considered a sort of gustnado. The line between these various types of whirlwind is not easy to draw!

Gustnadoes can also form at the rear-flank downdraught of a supercell thunderstorm, in the very quadrant where tornadoes of the mesocyclonic varieties have their birth. Gustnadoes forming here might be mistaken for weak tornadoes but, more importantly, if they form too close to the updraft/downdraught interface they are in danger of being drawn into the downdraught and spun up into the real thing. One again, the line between these classifications is crossed!

Steam Devils

In animate nature, it is not uncommon to find a harmless animal imitating the appearance of something far more sinister. In shallow pools along the Great Barrier Reef of Queensland, to take just one example, there is a type of sea slug that at first glance looks just like the deadly sea snake, but in reality is about the most innocuous creature that anyone could imagine. The reason for this mimicry is a very practical one; to avoid being eaten by predators! Yet, there is an odd parallel in inanimate nature as well, which can be nothing other than pure coincidence. This is the *steam devil*; truly the mildest of atmospheric vortices that manages to mimic the most violent.

Steam devils are rotating updrafts of steam which typically form in such environs as the cooling towers of electricity generators, geothermal vents, hot springs, in the clouds of steam above hot asphalt that has been recently sprayed or rained upon and also in the cold mist arising from frigid air spreading over the surface of a warmer body of water. Under such conditions as these, a type of low-level mist known as *sea smoke* forms and can be whirled into eddies, some of which assume the shape of long and narrow tubes of mist that superficially resemble the funnel clouds of waterspouts or tornadoes. Yet, although they are made visible by condensation of water droplets, the process causing this condensation

Fig. 5.10. Steam Devil (loop-like formation below center) (Credit: Mbzl/ Mbzl gallery/Picture Gallery Atmospheric Optics 3. Mila Zincova)

is quite different from that of a true funnel cloud (i.e. the cooling effect of a sudden reduction of air pressure) and the similarity is only coincidental.

Steam devils that assume very long and slender forms are thought by some to have given rise to stories of the Daughters of Nereus who are said to dance on the ocean waves and come to the rescue of endangered mariners. As the legend of the Mermaid is believed to have originated from sightings of Dugongs and Manatees draped in flowing "tresses" of kelp (and whose faces were presumably well hidden!) just so these slender feminine fingers of mist may have provided the raw material for this other endearing legend of the seas.

By the way, it might be thought that the vortices forming in cold mist of the sea smoke type might have acquired the name *mist devil* to differentiate them from the vortices of hot steam over geothermal power plants and volcanic lava flows, but apparently "steam devil" suffices for both. Perhaps there are as many varieties of "devil" as we can handle!

"Explosion Tornadoes"

Tornado-like whirlwinds are sometimes spawned by volcanic eruptions and other large explosions. They have apparently been noticed dangling down from the mushroom cloud of a nuclear bomb blast; a sight which we all hope and pray will never be seen again. Presumably, the impacts of giant meteorites have also given rise to tornadoes of this type and we may wonder if a parade of such whirling columns of cloud and smoke paraded across the face of land and ocean following the mighty impact that brought the Cretaceous Era to a close. Like the tornadoes spawned by nuclear bombs, these are twisters that we never want to see again. Fortunately, as automated sky patrols reach fainter and fainter magnitudes and cover increasing areas of sky in decreasing periods of time, we should become better armed against the disasters liable to create them.

Project 15: The Bathtub Vortex

Vortices occur in fluids other than air of course. Every time you let water out of a basin or bath, you see an example of a water vortex and you were probably told as a child that these spin in different directions depending on which hemisphere you are in. In the northern hemisphere, they spin clockwise (you were probably told) and in the southern, anticlockwise. But is there any truth in this story or is it just another urban legend?

   In theory, Coriolis force should determine the direction of spin of a water vortex, just as it does a whirling column of air. Yet, we know that smaller whirlwinds can spin either way. Mesocyclonic tornadoes are principally cyclonic, but they derive their spin from the larger mesocyclone that gives them birth and, needless to say, a bathtub or washbasin is miniscule in comparison with a rotating thunderstorm! Why should Coriolis force have any effect on so small a system?

(continued)

Project 15: (continued)

When scientists applied a little math to the subject, it seemed pretty clear that, for systems as small as baths and basins, Coriolis force was overwhelmed by other effects, such as the direction in which water entered the bath/basin and so forth. Studies of bathtub vortices in both hemispheres appeared to confirm this, with no directional preference in either hemisphere.

And yet! One experiment involving scientists in USA and Australia found that there was a preference in the direction of the vortices, provided that the water in the container was left undisturbed for several days until the "memory" of all "local" vortex effects faded. Apparently, a container of water holds this "memory" for much longer than we might otherwise imagine, but if the water was released through a small hole in the bottom of the container after remaining still for several days, the resulting vortex behaved in the theoretically correct way; anticlockwise in the USA and clockwise in Australia.

You might like to try this yourself. A bath or basin may not be the best vessel for the water, as reaching your hand into the water to extract the plug will probably add too much disturbance to the system to give an accurate result. A better vessel would be a large and deep tray with a valve underneath, that could be emptied without the need to touch or otherwise disturb the water. Set up the tray somewhere where it will not be disturbed, fill with water and allow to stand for 4 or 5 days. Then open the valve and note the direction of spin. Do this several times a statistically significant result emerges. If possible, persuade a Facebook™ friend in the other hemisphere to repeat the same experiment and then compare results.

Because Coriolis force increases with distance from the equator, this experiment should theoretically yield more obvious results if performed at higher latitudes. It would be interesting to see if several people living at a variety of latitudes (north and south of the equator) could perform the experiment according to the same specifications and compare results. If Coriolis force is really a factor, the results should be clearer for high latitude experiments but not as definitive for those relatively close to the equator.

Oddities of the Storm

There are about as many tornado stories as there are fishermen's' tales of the fish that got away and it is very often just as difficult to sort out the truth from stories that may charitably be said to have grown a little in the recounting. Yet, it is also true that some tales which on the face of it appeared impossible, turn out to be accurate after all.

Is There Something Fishy About This Rain?

One of the recurring stories involves fish and other marine creatures falling from the sky. Oddly, these are not necessarily told as *tornado* stories per se. In fact, many (maybe even most) of the fish-fall accounts do not involve the sighting of a tornado or other whirlwind. The tornado connection is in the explanation. How else could fish be taken up into the air unless they were lifted there by a strong atmospheric vortex such as a tornado (more likely in the form of a waterspout) or strong wind devil?

The weird part is, many of these fish falls occur relatively far from water, are often very concentrated and involve just one species of fish. It is as if a school of fish got taken aloft, carried a considerable distance and then dumped in the one spot!

It would be pointless to try to catalogue all the accounts of this phenomenon, but a few samples will show just how odd some of these events can really be.

As long ago as 1883, on December 15 of that year, a shower of live perch landed in Airdrie in Scotland. A certain John Stewart, reporting this phenomenon second hand to the journal *Knowledge*, perceptively remarked that perch do not normally swim close to the surface of the water, making them less vulnerable than certain other species to being taken aloft in a whirlwind. He concludes that, if a whirlwind was indeed the culprit, it must have been a very strong one. Strangely, this event was said to have taken place in the morning when (especially in winter time in Scotland!) one would not expect strong surface heating to have taken place. Was a tornado spawned by a winter cold front? What was the weather like at the time? The brief report did not say.

Moving to Australia and forward to 1924, the small inland town of Gulargambone in New South Wales was reportedly hit by a rain of fresh water fish, with "hundreds of small gudgeons" found in gutters all over to town. This was said to have taken place during heavy rain and was most probably caused by a tornado crossing the river, although there do not appear to have been any tornado reports (but then, maybe everyone was inside sheltering from the rain!). Other Australian reports include a rain of long marine worms at Randwick (suburb of Sydney) in 1927, three rains of fish at Killarney Station – some 200 miles (320 km) from the sea! – in February 1974, a further gudgeon fall in 1901 (this time at Warwick), yet another in suburban Brisbane in 1959 and another at Quirindi on 12 November 1913. Tornadoes were reported in the area on the day of the Quirindi event and the rain of fish was specifically reported as having come from a "tornado-type cloud." That time at least, the culprit seems to have been caught in the act. Not so satisfactory though was an incident in 1973, when a householder on the north coast of New South Wales awoke in the morning to find fish all over the roof of his home. Apparently, no other nearby house was affected and no nocturnal tornado reported!

Falls of frogs, lizards, sea shells and all many of small creatures have been reported for hundreds of years and from locations all over the world. Even stranger things have tumbled from the sky from time to time. Rains of nuts have allegedly occurred but probably the strangest report of all involved over 1,000 sixteenth-century coins that were said to have tumbled down into Gorky in what was then the USSR on 17 June 1940. This event was reportedly witnessed by a group of very interested schoolboys!

Although we might be inclined to put this tale into the "tall story" class, Russian meteorologist D. V. Nalivikin thinks that there is quite a rational explanation. Mind you, for a "rational" explanation, Nalivikin's hypothesis does sound a little weird, but that is not necessarily a problem. Nature *can* be weird at times!

For a start, he confirms that there *was* a tornado reported at the time and that the coins were said to have fallen from the parent thundercloud. He then draws attention to the existence in all vortex systems (including tornadoes) of inhomogeneities in the form of "jets" or "flows of air." These are, in effect, wind gusts of remarkable compactness. According to Nalivikin, these "jets" can be so compact and powerful as to act like air shovels capable of scooping up clusters of objects from the surface or even excavating them from a shallow depth, and carrying them in a compact group for considerable distances. On this view, the antique coins were literally scooped up from beneath the surface soil by a powerful jet of air which carried them as a compact cluster into the thundercloud and from there across a more or less large tract of countryside, finally depositing them at the feet of some amazed schoolboys.

This process may also explain why falls of the same kind of object are so frequently reported. That is to say, we might logically think that a tornado taking up fish from a river (say) would also pull up clumps of water weed, floating tree branches and samples of the usual detritus associated with rivers. Yet, we have report after report of just *one* kind of fish falling. No weed, no river gravel. Not even different species of fish. How can a tornado or other type of whirlwind be so selective? Surely there is something very weird afoot here!

Interesting certainly, but maybe not as weird as it looks. Imagine a jet within the tornadic vortex hitting a river and scooping up whatever happened to be at that place in the water; small fish, large fish, sand, gravel ... This mixture is taken up and held compact by the jet as it swirls upward in the vortex. Yet, before long the lightest objects will probably leave the cluster, blown out into the general circulation of the tornado. The heaviest objects will also fall away, leaving a nice compact cluster naturally sorted by weight. Moreover, the sorting may not be quite as neat as it appears. If someone finds his back yard littered with fish, the presence of pieces of wood and some gravel of about the same weight as these more conspicuous visitors would probably go unnoticed.

Sometimes very unattractive things fall from the sky. In Brazil on 27 August 1968, spongy meat and drops of blood fell from a clear sky over an area equivalent to several home properties. The clear sky would seem to rule out tornadoes in this instance. But did a powerful wind devil passed over a nearby abattoir ... ?

Pranks of the Tonado

As we said earlier, many tornado stories seem just about as fantastic as the much maligned fish story. For years, apparently credible reports of chickens having been denuded of feathers, pieces of straw and even the soft petals of flowers driven into tree trunks and the like were given as evidence of winds of truly fantastic velocity, even approaching the speed of sound. In some cases, the original sources of these types of stories have proven as elusive as the rainbow and are probably simply urban legends that have grown over the years. Other apparently amazing feats of the wind have been found to occur at slower velocities than had hitherto been thought possible. The point is, wind per se is not the only factor to be considered. Greatly reduced atmospheric pressure must be factored into the equation as well. An example of this is probably the extraction of the rather hefty boulder from the ground of virgin forest by the same tornado that the writer experienced as a child and with which this chapter opened. Undoubtedly, the sudden reduction in pressure at the ground surface caused air trapped in the soil beneath the rock to exert upward pressure which tended to "pop" the rock like a cork. A wind-speed estimate based solely on the force required to "blow" a boulder out of the ground – a boulder that was set fast in the soil since time immemorial – would certainly yield values far too high. Reduced air pressure also plays a role in the denuding of chickens ... and in the denuding of human beings as well! Cases have been reported of shoes being "blown" off the feet and the body of a man killed in a severe tornado at Sandon in Victoria (Australia) on 13 November, 1976 was said to have been naked when found. He was one of two middle aged people whose vehicle became airborne after being caught in the tornado. Sadly, both victims were pulled from the vehicle by the force of the storm, neither surviving their ordeal. The stripping of the gentleman's clothes was the final indignity of this tragic saga.

A similar, though fortunately neither so tragic nor embarrassing, instance of this reduced air-pressure effect was reported in association with the Buladelah (New South Wales) tornado of New Year's Day 1970. Fortunately, this storm expended its fury on virgin forest land, coming near but not touching the township of Buladelah itself. By all estimates of the tree damage left in its wake, it was a very severe storm and may have rated a full 5 on the Enhanced Fujita Scale. But as well as wreaking havoc on the forest, the tornado also performed a rather unusual feat. It sucked a wallet from a man's trouser pocket (a tornadic pickpocket! *Oliver Twister* perhaps?)! The gentleman concerned was outside the tornado itself (otherwise, a lost wallet would have been the least of his worries!)

but the surrounding air was being drawn up into the vortex with such force that the wallet was pulled right out of his pocket!

On the other hand, in the midst of these stories of almost incredible feats of destructive power, we find the occasional story of surprising gentleness. A typical example was noted in a tornado which bust upon Empire Bay, not far from the writer's home, on 10 August, 1984. This brief but intense tornado managed, in the course of between 15 and 30 s as estimated by eyewitnesses, to totally unroof 12 houses (sending the roofs spinning high into the air), partially unroof three more, flip over two trailers, take the second story off a two-storied brick veneer house, collapse the walls of three more houses, flatten a garage, damage an apartment building, lift a 1-t boat from its trailer, carrying it at least 60 ft and hurl a heavy iron girder tens of feet through the air. Yet, in all of this fury, a fragile and expensive glass lamp fitting was found undamaged on a swath of vacant land, far from the home from which the twister had snatched it away. Another tale, about whose truth I will not speculate tells of a man being taken up into the vortex of a tornado and, while flailing his arms around in mid-air, managed to take hold of the tail of an airborne horse. Later, the waning vortex delivered him down gently – with his hat in one hand and a tuft of horse hair in the other!

Whether this last tale is true or tall, the fact remains that a twister can at times be remarkably "gentle" in that people and fragile objects taken to considerably altitudes in the vortex are placed down to earth with remarkable tenderness. Of course, there are also accounts (like the Sandon fatalities) where this is certainly not so, but the fact that it should occur *at all* is wonder enough. The reason for these gentle touchdowns is the more or less gradual way in which the updraft weakens away from the centre of the tornado. Someone or something lofted by the central updraft may – if they are fortunate – find themselves sinking only gradually as they are whirled outward from the centre and into regions of progressively decreasing updraft strength.

Zones of Silence?

Tornadoes are notoriously noisy phenomena and a twister can be heard miles away heralding its approach by a constant dull roaring sound that builds to something like the scream of a jet engine as it comes close. Yet, there is some evidence for peculiar "zones of silence" directly beneath some tornadoes as they lift from the ground and, technically, become funnel clouds rather than tornadoes in the full sense. Evidence for this phenomenon is not strong in the sense that there are very few accounts available, but it is also true to say that not many people have been placed in the unenviable position of having a low funnel cloud poised directly above their heads and, of those that have, most were probably concentrating on something other than sound effects. Yet, there is one seemingly very credible case on record; a report by retired US Army captain Roy S. Hall who found himself directly beneath a powerful tornadic vortex in Texas in May 1948.

Initially, this tornado was accompanied by the usual roaring, rumbling and throbbing sounds as it swept across the ground. Then, suddenly, all became silent. Hall not surprisingly supposed that the tornado had collapsed and dispersed and went to look outside. But the tornado had not disappeared. Instead, it had lifted off the ground and the great funnel hung menacingly directly over his house at treetop level. Apparently, there was no noticeable vortex at ground level at that moment, however the tornado touched down again very shortly afterward and resumed its path of destruction. Whatever had caused the cessation of noise, it was not related to a weakening of the storm's fury.

It should be noted that several reports of other people caught directly beneath a lifted tornado have not mentioned a zone of silence. The famous account of Will Keller of Kansas who looked directly up through the center of a funnel on 22 July, 1928 implied that the roaring and screaming of the vortex reached a crescendo at the moment the vortex passed overhead. Furthermore, there are accounts of funnel clouds that, while not touching down as tornadoes, nevertheless become audible like "the buzzing of a thousand bees" relatively high in the atmosphere. The zone of silence remains just another mystery of our atmosphere.

Cloudbursts and "Hydrometeorites"

In most familiar speech, a "cloudburst" is simply a colloquial term for an extremely heavy, and usually rather brief, rainstorm.

As such, there is nothing strange about it. But there are occasional reports of "cloudbursts" that seem to be of an altogether different nature and appear in some way related to tornadoes. Thus, long ago on 31 May 1682, some 12–13 in. (about 600 mm) of rain was said to have been dumped on a very limited region of Oxford (England) in just 15 min. Even this pales in comparison with an alleged fall of about 52 in. (around 2,500 mm) at just one spot during the course of only a few minutes at Langtoff (England) in 1892. The amount of water in this phenomenal incident of point rainfall was derived from the erosion damaged that the event incurred.

In many of these incidents – including the Langtoff one – a "waterspout" is mentioned by witnesses. Typically though, the "waterspout" is seen over land and is spoken about as a literal tube of water. It is difficult to understand whether the witnesses imply something like a tornado or funnel cloud (which a "waterspout" over land superficially indicates) or whether they simply mean a dense shaft of rainfall. Sometimes quite ordinary rain shafts are referred to as "waterspouts" and it seems that some folk not familiar with meteorological terms believe that even true waterspouts are simply shafts of rainfall!

Yet, the tornado connection might be real in the opinion of British meteorologist G. T. Meaden. He suggests that a tornadic vortex should be capable of lifting and concentrating rain within the parent cloud into a watery mass that might best be described as a super raindrop, or even an entire shower of super raindrops. When the vortex dies away, or the weight of the mass of water becomes too heavy to be sustained by the updraft, or when the watery mass is spun outward from the most intense region of updraft, the collected raindrops fall to earth in a sudden, torrential, but very brief cascade.

There does indeed appear to be observational evidence of this. Sometime during 1974, a waterspout in the Pacific Ocean was seen to "break" at the juncture between the funnel cloud and the parent cloud base. According to witnesses, a "bulb-shaped object" formed near the juncture "and fell with a sudden rush into the sea." It was said to have hit the water with considerable force, sending up a great cloud of spray. The impression of this object was that of a huge drop of water. This report does not stand alone. An account of a far earlier waterspout also tells of a cloud-like fragment breaking away from either the funnel or the parent cloud and hitting the ocean "with a thundering roar."

If these falling bodies really were masses of water, it is very unlikely that they were formed from water lifted from the ocean surface. It is far more likely that they were formed from the cloud or rain by a process such as that hypothesized by Meaden. As such, the phenomenon is just as likely to occur in conjunction with terrestrial tornadoes as waterspouts. It may also occur in conjunction with funnel clouds, i.e. in situations where the vortex itself is not perceptible at ground level and may pass completely unnoticed. The lack of evidence of a tornado or waterspout in conjunction with an individual instance of intense point rainfall does not, therefore, rule out Meaden's hypothesis. The vortex may simply have remained aloft and unnoticed.

A similar process may even explain the curious accounts of falls of large quantities of ice that are reported from time to time. These days, reports of this nature tend to be explained away in terms of ice having formed and subsequently cracked away from the wings of aircraft or formed from aircraft waste water. In all likelihood, some occurrences are explicable in these ways, but this hardly does justice to the many eyewitness accounts from years before any aircraft flew. Selecting a few of these early cases, we find;

Late 1700s (exact date unknown). At Seringapatam in India, a block of ice described as being as big as an elephant and taking 3 days to melt fell from the sky.

July 24, 1818. In the Orkney Islands off Scotland, jagged pieces of ice up to "a foot long" fell from the sky. They reportedly emitted a sulfurous smell when melted.

October 1844. A block of ice "weighing 5 kg" fell at Cette in France.

March 16, 1860. During a snowstorm at Upper Wasdale in England, blocks of ice were said to have fallen from the clouds. No specific size is given, although they must have been very large as, after the storm, they were described as resembling a flock of sheep on the ground.

June 1881. A mass of ice 21 in. in circumference fell during a hailstorm in Iowa, USA.

In more recent times;

July 30, 1957. At Reading, Pennsylvania, a husband and wife witnessed the fall of a piece of ice some two feet in diameter.

September 8, 1958. Also in Pennsylvania, but this time at Chester, a large piece of ice came crashing down through the roof of a warehouse.

May 24, 1972. A building at Riverside, California, was hit by what reports called an "ice bomb." The chunk of ice, said to have weighed somewhere between 30 and 50 lb, crashed through the roof of a downtown law office building, causing several hundreds of dollars worth of damage. Although suspected of having fallen from an aircraft, a spokesperson for the Federal Aviation Administration said that he was not aware of any aircraft in the area at that time.

These reports are only the tip of the iceberg (no pun intended!) but they are sufficient to give a fair idea of the type of thing being witnessed. It is true that aircraft may have been responsible for one or more of the last three, but there is no real evidence of this and their similarity to the earlier cases where aircraft definitely were *not* involved suggests that some naturally occurring process was responsible for all the incidents listed here.

Sometimes the term "hydrometeors" is given to these ice falls. Although a more accurate term would be "hydrometeor*ites*", the term certainly betrays the thought of the people using it. They believe that these ice falls come from outer space. That suggestion was directly mooted as the possible explanation for a 30 lb lump that fell in Oklahoma in March 1982, and suggestions have been made that ice meteorites might be associated with comets.

Comets have been described as "dirty snowballs", but all research since the 1980s has discovered that the "dirty" part is not to be taken lightly. They would better be described as "frozen mudballs" and the thought that a slab of relatively clean ice dropping from the skies could have begun life as a fragment of one of these objects is highly improbable to put it mildly. Moreover, could we really expect an ice meteorite to survive passage through Earth's atmosphere, especially at the velocities associated with most cometary meteors?

A meteorological explanation is much more credible and something like Meaden's proposed explanation of cloudbursts looks like a good bet. That is to say, if the cloud in which the tornado/water-spout/funnel cloud was located consisted of *frozen* droplets, these might be gathered together into a very large "hailstone" in the manner that Meaden speculates that raindrops may be accumulated into an enormous raindrop. The occurrence of ice falls during hail or snow storms would appear especially amenable to this explanation, but even clear sky falls are not necessarily ruled out if the "clear" sky contains a few cumulus towers whose constituent drops are very nearly frozen. This is only speculation, but suppose a fair weather vortex or cold air funnel whirled together cloud droplets that were close to freezing in the parent cloud. May it not be possible that the cooling effect of the pressure drop within the vortex could just tip the scales sufficiently to cause the droplets to turn to ice crystals on contact, freezing together into a single, super-sized, hailstone? This may sound farfetched, and may well be as farfetched as it sounds, but I do believe it to be more credible than ice meteorites!

Fire, Electricity and Strange Odors!

The occurrence of electrical activity in tornadoes is something that had been noted as long ago as Lucretius (ca. 50 BC) and sometimes this gives rise to odd and interesting happenings. Strange luminous effects, glow discharges and even fire have been reported in association with tornadoes.

The mention of fire is, perhaps, the most disturbing of these effects and the one that many will find the least credible. Nevertheless, fire has been reported in conjunction with tornadoes from locations as diverse as America, Europe (especially France it seems) and Australia and from as long ago as circa 500 BC when the Greek sage Heraclitus of Ephesus wrote about "fiery waterspouts." Many of the reports show striking consistency even though remote from one another in both time and space. Thus;

Circa 396 AD. St. Augustine writes "A fiery cloud was seen in the east, small at first, then gradually as it came over the city it grew until the fire hung over the city in a terrible manner; a horrendous flame seemed to hang down and there was a smell of sulfur."

26 August, 1826, Carcassone (France). "At noon, massing clouds in the west. Then a thick black cloud suspended over a

field. Crackling and hissing and air rushed toward the cloud. A loud detonation and an enormous column of fire hovered over the field – throwing down walls and picking up and killing sheep. A strong sulfurous odor. Nobody said that it burned anything."

(The resemblance between these two widely separated accounts is especially striking.)

1869, Ashland (Tennesse). "A remarkably hot day ... a sort of whirlwind came along over the neighboring woods, taking up small branches and leaves and burning them in a sort of flaming cylinder." This account continues, saying that the whirlwind passed over a group of horses and singed their manes and tails, set fire to a field of corn and to the roof of a house before creating a cloud of steam after reaching a river.

18 July, 1881 Americus (Georgia). "At some distance from the town a small whirlwind, about 5 ft in diameter and sometimes 100 ft high, formed over a cornfield where it tore up the stalks by the roots and carried them with sand and other loose materials high into the air. The body of the whirling mass was of a vaporous formation and perfectly black, the center apparently illuminated by a fire and emitting a strange 'sulfurous vapor' that could be distinguished at a distance of about 300 yards, burning and sickening all who approached close enough to breathe it. Occasionally the cloud would divide into three minor ones [note the multi-vortex description here!] and as they came together again there would be a loud crash, accompanied by crackling sounds, when the whole mass would shoot upwards into the heavens."

Nineteenth century, Australia. Buried amongst the collections of astronomical and meteorological records of early New South Wales government astronomer H. C. Russell is a curious report of a witness who saw a tornado strike a tree and, upon investigating the damage after the twister had moved on, was surprised to find the tree smoldering. He felt sure that there had been no sign of fire prior to the tornado. Of course, the temptation here is to explain this away by assuming that a spark had been burning before the tornado struck and had merely been fanned by the wind. But then, we were not present and the witness who was present, insisted that no fire existed before the tornado!

It is difficult to believe that fire as such is present in tornado funnels. What is more probable is very intense electrical activity and it is this that both accounts for the visual appearance of fire in the clouds (as Augustine recounted) and the strong odor sometimes reported. Under suitable circumstances, it may set alight combustible material on the ground. On a few occasions, as we saw earlier, witnesses have been able to look directly upward into the funnel of an overhead tornado, and have specifically noted that a great deal of lightning is then visible within the funnel cloud itself. The famous example (mentioned above) of this is the experience of Kansas farmer Mr. Will Keller who had a tornado pass directly overhead on 22 July 1928. Looking upward into the centre of the funnel, he remarked that the central core was clear, but that short flashes of lightning constantly discharged from one side of the funnel's rim to the other and that the air was filled with such a strong gaseous odor, that it became hard to breathe.

Keller's mention of a strong odor recalls the ancient account by St. Augustine as well as the French and Georgian reports listed above. A strong pungent smell like gunpowder was also said to have hung in the air following a tornado at Manilla in New South Wales on 8 October 1932.

Luminous phenomena other than lightning in the strict sense, have been noted in association with tornadoes from time to time. Luminous waterspouts are not unknown, but the usual explanation of luminous aquatic organisms being drawn up into the vortex seems less likely than some sort of slow electrical discharge within the funnel cloud. Land tornadoes can be equally luminous, judging by a sighting at Champaign (Illinois) in 1942 of something like a searchlight beam extending from the base of a thunderstorm to the ground. As the witness watched this strange sight, the beam of light suddenly morphed into a tornado funnel!

An even more spectacular display was provided by a nocturnal tornado at Blackwell in Oklahoma in 1955. One eyewitness to this storm remarked that the light from the funnel was so bright that he could not look at it without being dazzled. He did note, however, that the chief source of the illumination appeared to be rotating around the funnel like the beam of a lighthouse.

On other occasions, bluish glows have been observed emanating from tornado funnels and illuminating the countryside and even ball-lightning-like globes have been reported floating out from the lower ends of funnel clouds. Indeed, the Condon Committee investigating reports of unidentified flying objects concluded that some nocturnal UFOs may well be luminous funnel clouds and even remarked that when typical UFO reports are compared with luminous tornado accounts, the latter turn out to be the strangest and most impressive!

Electrical fields have also been recorded in association with non-tornadic whirlwinds of the dust devil variety and Meaden even suggested that part of the buzzing sound associated with dust devils may be due to electricity rather than simply the sound of the wind. We might also recall from the previous chapter, the account of a desert whirlwind that began to generate bolts of lightning as it intensified up a canyon. Clearly, there is more to be learned about the electrical phenomena associated with tornadoes and other whirlwinds and whether these electrical fields play any role in intensifying the whirlwind itself or whether they are simply side effects of the atmospheric vortex. The capricious nature of these events does not make controlled study easy but I guess that just adds to the fun of the chase.

Other Worlds: Other Whirlwinds

Whirlwinds are not confined to planet Earth. Any planet with a sufficient atmosphere should spawn vortices of some sort, but to date the only extraterrestrial whirlwinds that have been observed are those that frolic across the deserts of Mars. The Martian rovers *Spirit*, near the Martian equator, and *Phoenix* at higher latitudes have each seen dust devils marching across the Martian plains and orbiting spacecraft have also noted many columns of dust spiraling upward from the planet's surface as well as numerous instances of "tracks" through the dust scoured out by earlier dust devils. Some of these tracks have been surprisingly long and wide and the term "Martian tornadoes" has been used to describe the vortices that must have carved them out of the sands. That term, though it describes the size of the whirlwinds well enough, could be misleading as true tornadoes do not happen on Mars. Presumably, when the planet had a thicker atmosphere capable of supporting storm clouds and rain, true tornadoes did occur, but those days are

Fig. 5.11. Dust devil on Mars, as imaged by Spirit Martian Rover, 27 June, 2005. Some of the Martian dust devils reach heights of 5 miles! (Credit: NASA)

long gone. What Mars has now is dust devils; lots of them coming in all different sizes, from small whirls to true giants. Not surprisingly, the larger ones seem confined to low latitudes, where ground temperatures are greater during the summer months.

The dust devils of Mars possibly play an important role in the chemistry of the planet's atmosphere. Together with the straight-line-wind dust storms that sometimes obscure large regions of the planet (even the entire globe in extreme instances), dust devils play a role in lofting dust into the thin Martian air. Dust storms may carry more dust aloft at one time, but they do not happen as frequently. Dust devils, on the other hand, appear to be always occurring somewhere on the planet.

One of the mysteries of Mars is the way methane gas disappears from the atmosphere. Another mystery is where the gas comes from in the first place, but given that *something* is generating a fair amount of methane on Mars, why is the gas not more prevalent in the atmosphere? What is destroying it?

Now, the flux of ultraviolet light on Mars is enough to break down molecules of methane quite rapidly in geologic terms. Nevertheless, the process should still take about 400 years or so. But that is still too tardy to account for the paucity of the gas in

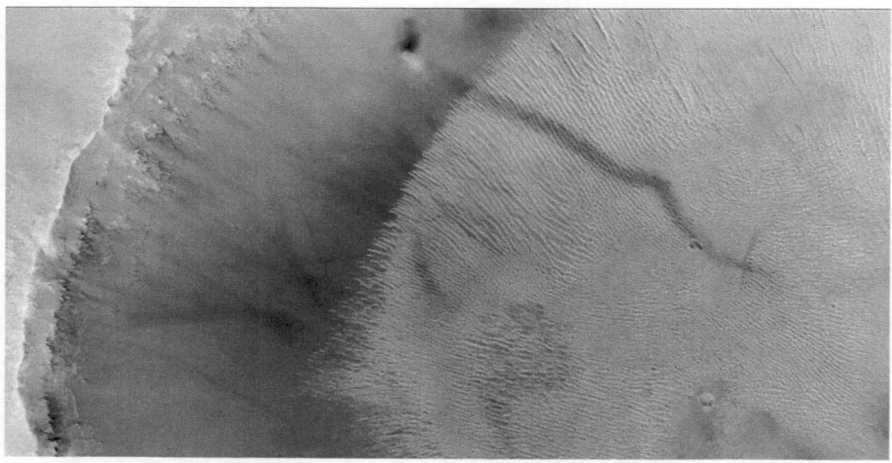

Fig. 5.12. A Martian dust devil casts a long shadow as it climbs out of a crater. Photograph by Mars Global Surveyoe (Credit: NASA)

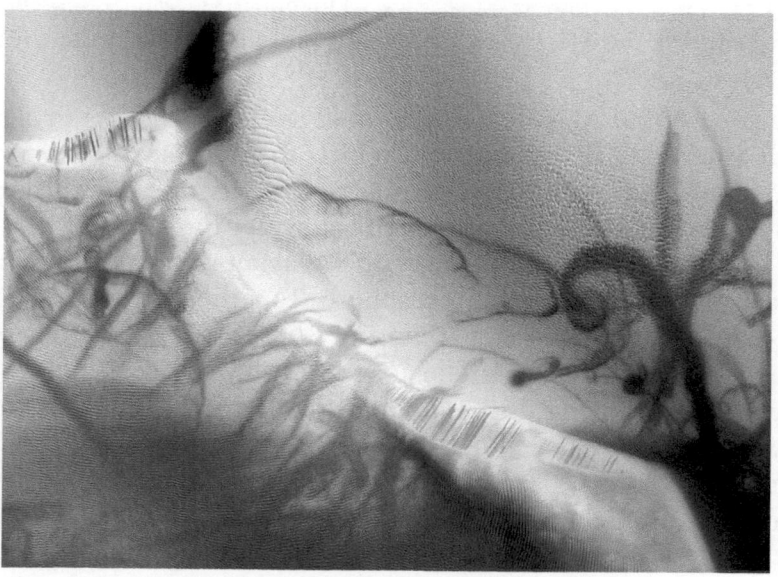

Fig. 5.13. Martian dust devils scour an intricate pattern of tracks in the dust as imaged by the HIRISE Camera on the Mars Reconnaissance Orbiter, 21 October, 2009 (Credit: NASA)

the Martian atmosphere, given the quantity that is known to be generated. If UV alone was responsible for breaking down methane molecules, there should be a much greater accumulation of the gas in the global atmosphere of Mars than we are finding.

The most likely explanation is that an oxidizing chemical present in the Martian dust is being lifted into the atmosphere and scrubbing it clean of methane molecules. And dust devils play a significant role in getting this oxidizing agent from surface to air. In this respect, Martian dust devils would appear to be having a more significant influence on the atmosphere of their home planet than terrestrial ones have on the air of Earth.

The Spouts of Kraken Mare?

For those who like a good sea shanty and also enjoy futuristic yarns of deep space exploration, here is a tip. Try seeking out *The Shoals of Kraken Mare* on the World Wide Web. The song was penned in 2010 and tells of a group of astronaut/sailors 100 years hence navigating the "waters" of Titan's great liquid hydrocarbon lake, Kraken Mare. The heroes of the song are met by 40 m waves, squalls of ethane sleet and run into real strife when their ship is holed by a shoal of tar.

But there was one thing that they did not have to endure. No mention is made of methane spouts!

Are there really methane spouts – analogues of the waterspouts seen on terrestrial oceans – on the liquid methane lakes of Titan?

We don't know at this time, but I cannot honestly think of a good reason why there should not be.

Titan is a slow rotator. It is also very cold. The slow spin of the moon means that atmospheric circulation is very different from that of Earth. For one thing, there is not the temperature gradient between equator and poles that we experience. To that extent, Titan is more of a low-temperature analogue of Venus than of Earth. The slow spin does not provide the kick for the large scale, or even mesoscale, atmospheric circulations that we experience on Earth. Mesoclclonic tornadoes probably do not occur on Titan.

On the other hand, convection clearly occurs in the cold atmosphere, as evidenced by the build-up of convective cumuliform methane clouds and the (thunderless) thunderstorm analogues that were observe by the Cassini spacecraft dumping torrents of methane rain at high latitudes. As on Earth, wind shear undoubtedly gives rise to horizontal atmospheric circulations which, on occasions, become caught up in the rising air currents of these developing cumulus towers and spun up into tornadic vortices of the "spout" type; landspouts over dry land and water-spout-analogues ("methanespouts") over the hydrocarbon lakes. Someday perhaps, robotic rovers will capture images of methane funnel clouds and swirling dust over the Titanian wastes or even vortices of liquid methane spray drifting across one of the moon's lakes. Maybe we will indeed see a spout on Kraken Mare!

Smaller convective vortices presumably occur as dust devils over the expansive deserts of Titan. The dry surface of the moon appears to be covered with an organic powder, believed to have been formed by the action of ultraviolet radiation on the methane rich atmosphere high above the moon's surface. This material floats downward as a steady fall of dust, finally settling on the surface where wind gathers it into dune fields somewhat reminiscent of the dunes of Earth's deserts. We can imagine dust devils swirling this material into the air in columns of organic dust.

The dust devils of Mars, though they can be quite intense, are not as menacing as they might appear due to that planet's very thin atmosphere. But the atmosphere of Titan is denser than Earth's. If intrepid astronauts ever do venture to this giant saturnian moon, dust devils and landspouts might present a real danger that needs to be assessed adequately and against which precautions might be required.

Twisters of the Gas Giants?

We know that Jupiter and Saturn breed thunderstorms of such dimensions as to dwarf planet Earth. We know that these huge balls of gas spin rapidly on their axes, sustaining winds of unimaginable ferocity and cyclonic eddies into which our entire planet could be immersed. Are tornadoes also spawned in these eddies and, if so, what must they be like?

Strictly speaking, if we hold to the "terrestrial" definition of "tornado", the answer to that question is "No." An atmospheric vortex is only defined as a "tornado" if it reaches ground level and, as there is no clearly defined solid surface on the giant planets, there is no "ground" for an atmospheric vortex to reach. However, in such unearthly situations, the definition of "tornado" can surely be extended and all we can say is that if tornadic vortices form in regions where the normal day-to-day straight line winds blow at several hundreds of miles per hour, our minds simply boggle at what such vortices might be like. And occur they do! All the ingredients of tornado formation – atmospheric eddies on a large scale, powerful updrafts from deeper within the planetary atmosphere and Coriolis force thanks to rapid planetary rotation – exist on a far larger scale than on Earth. Presumably, tornadoes exist there too, and on a commensurate scale.

We might also recall the excessively windy extra-solar planet HD 209458b from Chap. 4. That's the planet with the 4,400 miles per hour winds! If tornadoes spin up on that hot Jupiter, what would *their* wind strengths be??

Earth, even granted its tornadoes, dust devils, hurricanes and Arctic blizzards, begins to look like a very quiet place on which to spend our days …

Appendix: The Fujita and Enhanced Fujita Scales of Tornado Damage

The Fujita Scale was derived, by Tetsuya Fujita of the University of California, in 1971 to help researchers determine the strength of tornadic winds necessary to cause different levels of damage. Principally, it is a scale of wind *damage* rather than wind strength per se but from assessment of the degree of damage, wind velocities can be derived. The minimum wind speed for each step in the scale may be calculated by;

$V = 14.1(F + 2)^{1.5}$ miles per hour, where V is wind velocity and F is the scale value. The answer can be multiplied by 8/5 to convert to kilometers per hour.

The Fujita Scale is as follows;

F0 Weak Tornado (40–72 mph; 64–116 kph). Light damage. Signboards damages; some windows broken; twigs snapped of trees and shallow-rooted trees pushed over.

F1 Moderate Tornado (73–112 mph; 117–180 kph). Moving automobiles pushed off road; light trailers overturned. Outbuildings demolished and trees broken.

F2 Strong Tornado (113–158 mph; 181–253 kph). Roofs torn from frame houses; mobile homes destroyed and automobiles blown off roads. Large trees snapped or uprooted.

F3 Severe Tornado (159–207 mph; 254–332 kph). Walls torn from frame houses; locomotives derailed; automobiles lifted off ground. Most forest trees uprooted.

F4 Devastating Tornado (208–262 mph; 333–419 kph). Frame houses destroyed; bark stripped from trees; automobiles and locomotives thrown and rolled; trees in forest uprooted and carried some distance.

F5 Incredible Tornado (263–320 mph; 420–512 kph). Whole frame houses tossed from foundations; automobiles thrown through air and asphalt torn from roads. Trees debarked.

No tornado has been recorded as greater than F5, although the one in Oklahoma on May 3, 1999 packed wind strengths as high as 318 mph at its strongest; just shy of the "unbelievable" F6 rating. As there is no reason to think that this was the most violent tornado ever, it is very likely that tornadoes of F6 rating have occurred at rare intervals and it is not unlikely that some that have been rated toward the top end of the F5 category have actually fallen over the line into the low F6 category.

More recently, the scale has been updated by a team of U.S. meteorologists and wind engineers. This *Enhanced Fujita Scale* was implemented in The U.S. on 1st February, 2007. This enhanced scale is based upon the level of damage caused by three-second wind gusts. Wind speed within the gusts is estimated according to eight levels of damage to 28 types of structure, e.g. homes, automobiles, trees, towers and so forth. The Enhanced Scale is as follows;

EF0 65–85 mph (104–136 kph)
EF1 86–110 mph (137–176 kph)
EF2 111–135 mph (177–216 kph)
EF3 136–165 mph (217–264 kph)
EF4 166–200 mph (265–320 kph)
EF5 >200 mph

No value greater than 5 exists on the EF Scale, unlike the original F Scale which is, in theory at least, open ended.

6. Earth's Restless Envelope

The Ever Changing Clouds

Clouds are the bane of astronomers, yet we must admit that they can be things of great beauty. Perhaps familiarity breeds contempt, but just imagine how we would think about a billowing cumulus if such clouds were as rare as, say, meteor storms. People would be out in the streets admiring it! But in the real world, we might give it a passing glance and get on with the task in hand.

As we learned at school, clouds are divided into the basic broad classes of cumulus (towering clouds), cirrus (high clouds composed of ice crystals) and stratus (layered clouds). We were also taught that clouds bringing rain are classified as nimbus, but these come in two basic varieties, namely nimbostratus or layered rain clouds and cumulonimbus of towering rain-bearing clouds. The former are the rain clouds par excellence and the precipitation falling from them is typically in the form of small drops. This is what the weather forecast means when it predicts *rain*; small drops (though not *very* small drops) falling from nimbostratus clouds. Precipitation from cumulonimbus clouds typically takes the form of larger drops and falls over a limited region. This is still rain; such a rain event is what meteorologists refer to as a *shower* (or, if there is lightning present, a thunderstorm). A weather forecast predicting "rain and showers" is not repeating itself, as may at first sight be thought. It simply means that there will be more or less widespread rain from nimbostratus clouds and heavier dumps over limited areas from cumulonimbus clouds. If the cloud droplets are frozen, snow is experienced and if updrafts in cumulonimbus clouds are powerful enough, ice crystals in the upper regions of the clouds are balled together as hail.

When very small drops of rain fall to the ground, the precipitation is called *drizzle*. Light drizzle can fall from thick stratus clouds that do not quite qualify as nimbostratus.

D.A.J. Seargent, *Weird Weather: Tales of Astronomical and Atmospheric Anomalies*, Astronomers' Universe, DOI 10.1007/978-1-4614-3070-4_6,
© Springer Science+Business Media New York 2012

Clouds are also classified according to their altitude. The lowest of them all rests at ground level and is known as *fog* (or *mist*, if it is thin enough). The next level are *low clouds*; ordinary *stratus*, *nimbostratus*, *fractostratus* ("scud" as it is often termed) and fair weather *cumulus*. Above these are the *middle level clouds*; *altostratus* (the dull grey sheets which cover the sky and give rise to gloomy days) and *altocumulus* (patchy clouds which can give the sky an almost quilt-like appearance, sometimes called "mackerel sky" although this term is more correctly given to another form of cloud as we shall see shortly). Above these again are the *high clouds*; *cirrus* (wispy, feathery, trails commonly called "mares tails"), *cirrostratus* (icy sheets that can cover the entire sky and give rise to spectacular haloes around the Sun and Moon, as well as producing sun dogs and other displays of iridescence) and *cirrocumulus* (little puffy ice clouds looking like ripples in the sky. Large patches of these clouds are the true "mackerel sky." The individual ripples are quite a deal smaller than the altocumulus patterns to which this name is also sometimes applied, as already mentioned).

Towering *cumulonimbus* clouds go through all of these layers. Their bases can be very low, even touching the ground at times, but their tops may reach the altitude of the highest cirrus. When mature, with rain falling from their bases, the icy tops of large cumulonimbus *glaciate* or spread out into a special kind of cirrus cloud known as an *anvil* cirrus. When this formation is seen, rain is falling from the base of the cloud and its decay has already begun.

Far above any of these types of cloud – at the very edge of outer space – strangely beautiful waves and billows are sometimes seen from the higher latitudes of Earth. Clouds of this type form at altitudes of around 50 miles (80 km) above sea level. This is about the height at which most meteors are seen and is close to the so-called *Karman Line* (the altitude above which the atmosphere is too thin for aeronautical purposes) that is generally considered to be – for all practical purposes – the boundary between the atmosphere and outer space. So tenuous are these high altitude clouds that they can only be observed while still illuminated by sunlight, but after darkness has fallen on the Earth below. Because of their nocturnal appearance, these strange clouds have been given the name of *noctilucent* (night-shining) clouds. We shall speak more about these beautiful clouds later in this chapter.

Oddities Amongst the Clouds

Clouds may be the most common inhabitants of our skies, but that does not give them immunity from reports of the strange and the bizarre. Odd shapes and strange formations are not uncommon, but these rarely challenge our knowledge of the processes causing cloud formation or the processes that occur within them. Yet, buried amongst descriptions of cloud formations some real oddities lurk, of which the following examples are, perhaps, the strangest of them all.

Sometimes, clouds of a distinctly unusual nature are reported and it is not always easy to explain just what has been seen in these instances.

Luminous and Noisy Clouds

Consider, for example, the following report of a luminous cumulus cloud by J. Zeleny and published in *Science* for January 15, 1932. The incident had, according to Mr. Zeleny, taken place "some 35 years ago" which would have made it in the mid to late 1890s. The observation took place during an otherwise clear summer night at Hutchinson in Minnesota. Zeleny writes;

> The cloud had a horizontal diameter of about a third of a mile and a thickness of about one fourth of that distance. It rose majestically from the eastern horizon, shone with a uniform, steady, vivid, whitish light and passed directly over the town. When the cloud was overhead a great shower of insects descended to earth covering the ground all around to the number of about 50 to 100 per square foot. These insects proved to be a species of hemiptera and were non-luminous. They had apparently been induced to take wing by the bright object in the sky.

The author of this account goes on to raise, and by and large dismiss, a number of speculative suggestions as to what this strangely luminous cloud may have been. He dismisses simple reflection of city lights as being totally inadequate to account for the intensity of the illumination. He likewise rejects the possibility of reflected sunlight, as the event took place too long after twilight for that to be possible. He admits that he cannot recall if the Moon rose later that evening (although he is sure that there was no Moon visible at the time of the cloud) and is willing to entertain the possibility that, if the Moon did appear later that night, reflection of light from "a bright Moon below the horizon" might be a possible explanation. He also raises the possibility that "the cloud consisted of a mass of organic vapor that was slowly oxidizing, being in fact a case of an extended will-o-the-wisp", although he dismisses this hypothesis as also being "unlikely."

Personally, the present writer also finds the reflected moonlight hypothesis "unlikely." Apart from the doubt as to whether the Moon was even visible later that night, the intensity of the cloud's luminosity and the unusual feature of insects having been attracted to its light seems to place it in a different category to the usual moonlight illumination of clouds. Indeed, the observation of insects would appear to suggest that the cloud was at a relatively low altitude (which, incidentally, may mean that Zeleny's estimate of its dimensions might be a little exaggerated). A low altitude for this cloud makes illumination by a Moon below the local horizon even less likely.

Another possibility not mentioned by the witness is an electrical discharge similar to St. Elmo's Fire or (albeit on a reduced scale) the glows occasionally seen on the tops of thunderclouds. Apparently, no lightning was observed however and this suggestion must remain as speculative as any other.

This author, in the same article as the above, also recounted his sighting of a more recent and, in its own way, equally unusual cloud phenomenon. Not this time a luminous cloud, but a peculiarly noisy one!

This event occurred in the early morning of a day in late July1931 near Cache Lake in Ontario. Zeleny recalls that the morning was "chilly ... and the sky was completely overcast with clouds." He writes that,

> My attention was attracted by a rumbling sound coming from the west, such as heralds the approach of a heavy thunderstorm. As I watched, a very long, low, narrow, tenuous cloud, resembling a squall cloud, appeared above the trees on the

opposite shore, moving at right angles to its length. The continuous, rumbling noise, now grown remarkably loud, seemed to come unmistakably from this cloud, whose cross-sectional diameter was only about 200 feet. The cloud passed overhead eastward and was not followed by the expected rain storm. The cloud apparently marked the meeting place of two oppositely directed currents of air that differed in temperature. It seemed almost incredible, however, that so much sound could have arisen from the agitated air alone, and yet this seems to be the only plausible explanation of its origin. I steadfastly looked for small lightning flashes in the cloud and saw none, although they would have had to come in rapid succession to produce the persistent sound that was heard. The noise could not have come from the rattle of hail because the cross-section of the cloud was too small for hail formation.

The tell-tale statement here is "the cloud apparently marked the meeting place of two oppositely directed currents of air ..." Although there is no mention of rotation of the cloud, it appears most likely that this was a remarkable instance of a horizontal funnel cloud formed in a vortex triggered by the convergence of the opposing air currents. Horizontal funnel clouds are reported from time to time, although this seems to have been a very extreme instance of this phenomenon. In lieu of a better explanation however, it does appear the most probable one.

Not so satisfactorily (?) explained were the strange reports of "flapping, crinkling, crackling clouds" that drifted over Jacksonville Beach, Florida, in early February 1969. The sounds were reportedly heard by hundreds of people, including the Police Chief and Captain Harold Bryan. Following orders from the Chief, Captain Bryan followed the cloud to the edge of the ocean, where it dissipated. The crowd that had gathered started to disperse after the cloud's disappearance, but then a second cloud formed and repeated its predecessor's noisy performance. Eventually, it too dissipated over the Atlantic waters.

Witnesses of these two strange clouds described the sounds as being like "someone rattling cellophane", "like a big bird flapping its wings" or "like huge sheets of wax paper being crumpled." Neither the officials at the Mayport Naval Air Station nor the U.S. Weather Bureau in Miami could offer any explanation for the mysterious sounds.

Neither can the present writer, although I may just draw attention to a certain similarity between the descriptions just quoted and the reports of electrophonic sounds accompanying aurora, bright meteors and lightning flashes mentioned elsewhere in this book. What – if any – significance this may have is, however, beyond me!

Cylindrical and Circular Clouds

The Gulf of Carpentaria coast of northern Australia sets the stage for one of the Atmosphere's strangest cloud formations; the *Morning Glory*. A regular event in that part of the world, the MG does not, as far as we are aware, have an exact double anywhere else on Earth, although spectacular cylindrical clouds are occasionally reported elsewhere which bear at least some superficial resemblance to it.

The MG appears as an arch of cloud stretching right across the sky. It is surprisingly narrow; ranging in thickness from little more than 320 ft to about 650 ft (100–200 m) and about 1,600 ft (500 m) above the ground, although there are times when it occurs at just one tenth of that altitude. By contrast, its length is most impressive – from aircraft, it has been measured at 75 miles (120 km)! According to witnesses "Both the bases and the tops (of the cloud cylinder) are quite regular and well defined but the tops are inclined ahead of the bases. This gives a rolling effect, but the cloud is not always observed to be actually rotating ... double Morning Glories are commonly reported and as many as seven successive rolls have been observed. They are usually orientated north-north-west to south-south-east and move from east to west at speeds which range from [22 to 56 miles (36–90 km) per hour]. In some cases of multiple rolls the whole entourage may take more than an hour to pass. The clouds do not usually produce precipitation but a fine mist sometimes occurs with multiple rolls." The clouds are, however, associated with a sudden wind squall.

The phenomenon is most probably explained in terms of atmospheric solitary waves or undular bores, analogous to the tidal bores occurring at water surfaces.

Other unusual cloud formations worthy of mention are the occasional instances of circular clouds with hollow central regions; in other words, *rings* of cloud. A fine specimen was reported back in December 1922 at Biggin Hill Aerodrome in England. Described as having the form of a gigantic smoke-ring, this nocturnal cloud was some 60° across and thin enough toward the center for stars to be occasionally visible through it.

Another interesting example occurred on the moonlit night of February 1, 1958 as reported by Captain Digby Jones of the *Archilles*. This ring of cloud passed directly above the ship and was observed to rotate in a clockwise direction around its clear center. As it passed over the ship, the temperature dropped, the atmospheric pressure fell and the wind veered completely around. All returned to normal as soon as the cloud passed by.

From the last account, it would seem that at least some of these ring clouds mark the periphery of small (approximately mile-sized) cyclonic vortices that are not "officially" recognized with a specific name. The phenomenon is rare, but any observations of these "cloud rings" – especially if coupled with barometric data – could prove most instructive about the smaller and weaker circulations of our atmosphere.

Do Clouds Fall Victim to the Full Moon?

Astronomers might be excused for thinking so! How often do clear, clean, skies spoiled for deep sky observing by a brilliant Moon turn into gloomy palls once the Moon gets out of the way? Or is that just the perversity of human nature recalling the frustration of these nights while conveniently forgetting the clarity of all those dark skies when our satellite is absent?

On a more serious note however, some astronomers have truly believed that the full Moon does have a tendency to disperse clouds. Humboldt spoke of this belief as being well known amongst the population of South America. Approaching the issue from, as it were, the opposite direction, D.F. Arago held rain to be more prevalent around the times of new Moon, suggesting that the relative lack of rain near the full phase is due to the lack of cloudiness then. Even as notable a pioneer of modern astronomy as Sir John Herschel upheld this opinion and offered an explanation as to why it should be so. According to Herschel, solar heat radiated back to Earth by the full Moon has a drying effect on the atmosphere which, in turn, tends to restrict the formation of clouds. If true, this would mean that bright moonlight nights should also be warmer (other things being equal) than moonless ones. As far as I am aware, no claims to that effect were made.

There seems little doubt that this alleged effect, albeit supported by the likes of Herschel and Humboldt, is nothing more than a popular superstition. The degree of heat reaching our atmosphere after reflection off the lunar surface is altogether inadequate to do the job that Herschel required and, in any case, genuine observational data of the type amenable to statistics (and not simply folk tales and rumors) have not confirmed any detectable effect of the full Moon on clouds.

On a slightly different topic, there was a deal of speculation in the early 1960s as to the possibility of meteor dust acting as a source of nuclei for the formation of cloud droplets. There was a suggestion that the Moon may sometimes deflect meteor dust into our atmosphere, but the most efficient depositors of dust were thought to be meteor showers themselves. Attempts were made to correlate rainy periods with the occurrence of the major annual meteor showers, but nothing very convincing was to come forth.

Interestingly, the writer recalls an incident when an astronomical society to which I then belonged attempted to use the meteor hypothesis to find a suitable date for their next open night. A series of possibilities ranging from "good" through to "bad" were worked out but, unaware of this, the committee responsible for arranging open nights had already decided on a date. You guessed it! The night with the worst prospects! Not only did the meteor hypothesis give the night the "bad" forecast, but even the official weather forecast for that day predicted showers and cloud. The result; a cool and clear night, a beautiful quarter Moon and a happy observing time by all!!

Shining Clouds of the Nighttime

Earlier in this chapter, we mentioned strange types of clouds forming at the very edge of outer space, so high above the Earth's surface that they remain illuminated by sunlight long after darkness has fallen at ground level. These are the beautiful *noctilucent* or *night-shining* clouds. Poles apart from the star obscuring nocturnal clouds that are the bane of astronomers, these beautiful clouds present as fine wisps and folds of the finest gossamer delicately shining in an otherwise dark sky.

Lying at an altitude of around 50 miles, these clouds are far above Earth's weather systems and, indeed, most of its atmosphere. The altitude normally taken as the boundary of outer space (the so-called Karman Line mentioned earlier) is about 62 miles (100 km) above the surface, so any higher and these clouds could claim astronomical rather than meteorological status!

Strangely, no records of noctilucent clouds exist prior to 1885. Is it simply a coincidence that this date is just 2 years after the massive eruption of the Krakatoa volcano in Indonesia? Probably not, but the reason may not be the one that immediately springs to mind. Large volcanic eruptions may stimulate the formation of noctilucent clouds, but that direct effect was probably not the most important one in bringing about the clouds' discovery. More likely, the spectacular sunsets and other high-altitude coloration following in the wake of the Krakatoa event made people more aware of cloud-like phenomena catching the late rays of the Sun.

Fig. 6.1. Noctilucent clouds over Bargerveen, Drenthe, Netherlands, 23 June, 2009 (Credit: Hrald)

ISS017E011632

Fig. 6.2. Noctilucent clouds photographed by crew of International Space Station, 25 August, 2008 (Credit: NASA)

At the time, the clouds were thought to be due to lingering remnants of the dust shroud that had enveloped our planet following the dramatic eruption, but when they continued being reported years after all other effects attributed to volcanic dust had ceased, they became recognized as a largely unassociated phenomenon.

One of the first people to study noctilucent clouds was the German Otto Jesse, who also secured the first photograph of the phenomenon as early as 1887. He appears to have been the one to coin the term "noctilucent cloud." Following the volcanic eruption of 1883 but prior to the noctilucent cloud discovery in 1885, Jesse had been studying the unusual sunsets and other sky effects caused by Krakatoa's volcanic dust, but did not at that time observe anything similar to the clouds which he came to regularly note after 1885. We have his own statement that he is unlikely to have missed such a phenomenon, and from this he concludes that 1885 was indeed the year that they started appearing. It must be admitted however, that it is quite a jump from saying that the "wave" of noctilucent clouds, which he observed, began abruptly in 1885 to the broader conclusion that noctilucent clouds per se did not exist before that date. Be that as it may, the appearance of the clouds after 1885 certainly drew the attention of the staff of Berlin Observatory. Between 1887 and 1896, Jesse, together with colleagues W.J. Foerster and Stolze, continued systematic photographic observations from the Observatory and it was during the progress of this research that the great altitude of the clouds was first determined via the process of triangulation.

If the clouds are not composed of volcanic dust, what then are they made of? Some early speculation proposed meteor dust, but except for the brief dust trains of large fireballs, it is hard to see how such concentrated clouds of meteor dust could arise.

Recent research indicates that noctilucent clouds are composed of water ice in the form of tiny crystals not exceeding 100 nm in diameter. Yet, the meteoric association may still hold up to a point, as the fine dust particles of micrometeorites or cosmic dust may act as nuclei for the formation of these tiny crystals. Fine dust wafted up from the troposphere (the lowest atmospheric level where Earth's weather events take place) and maybe from volcanoes likely add to the mix as well and it is even possible that in the high atmospheric region where noctilucent clouds form (known officially as the *mesosphere*) tiny ice crystals might form directly from water vapor; something that does not happen at lower atmospheric levels.

The presence of water vapor at these extreme altitudes is not readily explained, and this difficulty probably lies behind the early suggestion that the clouds may be entirely composed of meteor dust. Suggestions for the water's origin range from transportation from the troposphere through gaps in the *tropopause* (the boundary dividing the turbulent *troposphere* from the serene *stratosphere*) to synthesis in the stratosphere via the reaction of methane with hydroxyl radicals. A very controversial suggestion relies on the frequent impact of tiny *minicomets*, about the size of houses, evaporating as they strike the upper regions of the atmosphere. It should be remarked that although such objects do exist (as the SOHO spacecraft has shown) there is no evidence that they occur in numbers vast enough to strike Earth except rarely. A cosmic source for the water vapor does not seem likely, although it is an interesting speculation.

Even so, it must not be thought that the atmosphere at these levels has a high content of water vapor. In truth, the water content is about one hundred millionth that of the air over the Sahara Desert!

What water vapor that is present there can only freeze into ice at temperatures colder than approximately -184°F (-120°C). These temperatures are mainly encountered over the poles and (paradoxically though this may seem) during the summer months. Noctilucent clouds are essentially a polar, or at least a high latitude, phenomenon and one mostly confined to the summer months. They are more frequent at high latitudes in the northern hemisphere and, although not completely unknown at very high southern latitudes, the southern clouds tend to be less bright and a little higher than their northern counterparts.

Interestingly, the brightness of noctilucent clouds varies with the solar cycle, being more intense during times of solar *minimum*. In this sense, they are opposite to auroral displays which are more intense during times of solar *maximum* or shortly thereafter. Amateur astronomers at high latitudes can therefore be glad that as auroral displays decrease, noctilucent cloud activity can be expected to intensify! This solar cycle dependence hinges on the increase in ultraviolet radiation from the more active Sun breaking apart molecules of water vapor in the high atmosphere and thereby reducing the amount available for noctilucent cloud formation. Oddly however, the lowest period for the clouds does not strictly coincide with solar maximum (or solar UV maximum) but seems to lag by about a year. In fact, the general waxing and waning of the clouds is out of step with the solar cycle by about one year throughout. This is a mystery which has yet to be solved.

Another mystery of the clouds is their high radar reflectivity within the frequency range of 50 MHz to 1.3 GHz. Paul Bellan of Caltech proposes that this high reflectivity may be caused by a thin film of sodium and iron coating the ice crystals composing the clouds. That would do the trick alright, but the problem then is to explain where the metals for this coating might come from. Bellan suggests that it may originate in the iron and sodium atoms stripped off the surfaces of incoming meteoroids. These atoms initially settle into a layer at a slightly higher altitude than the noctilucent clouds but might afterwards slowly drift downward where they could freeze onto the ice crystals of any clouds that happened to be present. Experiments demonstrate that sodium vapor does indeed freeze onto ice at the extremely low temperatures of the clouds and other observations find depletion in sodium and iron, at the levels where these metals accumulate, during times of noctilucent cloud activity. Nevertheless, although this hypothesis might appear the best bet at present, it is far from proven and remains controversial.

During 2010, an interesting experiment was carried out that provided some new and interesting insights into the noctilucent cloud phenomenon. It was also remarkable in demonstrating the extent to which amateur scientist/professional scientist co-operation can make real progress in our understanding of the natural world.

The experiment was the brainchild of Welsh amateur astronomer and aerial photographer John Rowlands who, in co-operation with atmospheric expert Professor Nicholas Mitchell of the University of Bath, sought correlations between noctilucent cloud formation and planetary wave events by coordinating mesospheric temperature readings of the VHF meteor radar transceiver in Arctic Sweden and visual noctilucent cloud observations from his home in Wales. Planetary waves are atmospheric waves that extend coherently around a circle of latitude, and can reach significant amplitudes in the stratosphere and mesosphere. They cause oscillations in temperatures and atmospheric winds having periods of a few days. In winter, planetary waves generated in the troposphere flood upwards filling the mesosphere, but in the summertime (when the experiment was conducted) strong westward winds in the stratosphere prevent the upward spread of planetary waves and in effect trap them at lower (tropospheric) altitudes At least, that is the fate of most of the waves. Those having very large amplitude and periods of around 2, 5 and 16 days are, however, observed to extend into the summertime mesosphere to heights similar to those of noctilucent clouds, causing temperatures at these altitudes to rise and fall as the waves pass. These waves are thought to be natural resonances of the atmosphere and are actually *generated* at these altitudes by instabilities in the planetary scale atmospheric circulation. As noctilucent clouds are essentially a summertime phenomenon, it is within this situation that they form and the chief aim of the experiment was to see how they are influenced by this special variety of planetary wave.

The experimenters found a moderately strong correlation between the extent of noctilucent cloud activity (essentially, how far the clouds reached above the local horizon at Rowlands' observing station at Anglesey in Wales) and the cooling phases of the 5-day planetary wave. They also found that other factors influenced the cloud displays as well – not surprisingly, their data tended to confirm that a slight rise in solar activity has a noticeable effect on noctilucent cloud formation in so far as the 2010 season did not provide the same level of displays as that experienced during the 2009 season when solar activity was very weak. Rowlands' results also supported anecdotal evidence that noctilucent clouds tend to be more frequent during the early morning hours. The researchers suggest that this is due to thermal atmospheric tides that recur on a day-to-day basis.

As happens so frequently in experiments, there was one anomalous result. The night of June 14/15 yielded a beautiful display of noctilucent clouds, even though it was out of phase with the planetary wave data from Sweden. The mystery was, however, solved by an examination of images of the sky taken on that night. These revealed the sky to be filled with streaks and billows of gravity waves, of the type responsible for banded airglows and neat rows of high cloud (see Chap. 3 of this book for an account of this phenomenon). Apparently the noctilucent cloud display of that night was triggered by gravity waves propagating to mesospheric altitudes from some localized disturbance in the lower atmosphere, not from planetary waves generated at the levels of the clouds themselves. Solving the riddle of this anomaly therefore yielded still further insight into the causes of the phenomenon. How often has this been experienced throughout the history of science?

Atmospheric Displays of Light and Shadow

The interplay of light, shadow and atmospheric conditions in general is responsible for some interesting, sometimes truly spectacular and occasionally even frightening visual phenomena. Refraction and reflection of light, twilight effects, sudden changes in atmospheric opacity, sightings of apparent objects that are simply not there and some peculiar illusions are all part and parcel of the strange effects of our atmosphere and skies. To these, we shall now turn.

Dogs, Bows, Pillars and Phony Suns

Sometimes, bright lights appear close to the Sun which, at their most intense, can give the impression of second (or third!) suns in the sky, albeit not fully as brilliant as the real one. With such names as *sun dogs, mock suns* or, more correctly *parhelia* these brilliant and spectacular apparitions have long caused wonder, as is evidenced by the following passage in *On the Republic*, written by Cicero between the years 54 and 51 BC;

> Be it so, said Tubero; and since you invite me to discussion, and present the opportunity, let us first examine, before any one else arrives, what can be the nature of the parhelion, or double sun, which was mentioned in the senate. Those that affirm they witnessed this prodigy are neither few nor unworthy of credit, so that there is more reason for investigation than incredulity.

Sun dogs form when the Sun shines through a layer of hexagonal ice crystals. This can happen anywhere in the world where there is a layer of high cirrus cloud, but in very cold conditions, low-altitude sun dogs can also be formed by a haze of ice crystals ("diamond dust") drifting around close to ground level. Hexagonal crystals act like little prisms refracting the Sun's light at an angle of 22°. If they are randomly orientated, the incident light is refracted in all directions away from the Sun, with a complete solar halo being the result. But if, as often happens, the crystals align vertically as they drift downward through the air, the incident sunlight is refracted horizontally and two rainbow-like images of the Sun appear on either side of the Sun at the same altitude. Two images, of course, presuppose that the sheet of cirrus cloud extends far enough on either side of the Sun. Needless to say, if it is absent on one side, only one bright spot appears.

Sun dogs are mostly seen when the Sun is relatively low in the heavens. At high elevations, its rays passing through the prismatic crystals are increasingly skewed and, if sun dogs happen at all, they are more displaced from the Sun. Sometimes solar haloes and sun dogs appear together, the latter appearing as bright jewels on either side of the ring. In these instances, both randomly-orientated and vertically-aligned crystals exist in significant numbers.

Sometimes, Sun dogs can appear quite elongated. In these examples, the colors are more clearly distinguishable. Yet, even sun dogs that assume the more regular shape are quite colorful, albeit not having such clearly defined bands as a normal rainbow. The side nearest to the Sun is at the red end of the spectrum, and colors grade out from orange through blue as we look further from the Sun, finally merging into the white of the solar halo, if one of these is also present.

A rare version of the sun dog is known as the *false sunrise*. This is in no way to be confused with the "false dawn" which is the old name given to the cone of Zodiacal Light that appears before the first glimmers of the true dawn, as we saw in Chap. 3. Likewise, the false sunrise here referred to is unrelated to the Novaya Zemlya mirage of the almost-risen Sun. This could be called *a* false sunrise, albeit one of a very different variety!

The phenomena of which we now speak is a sun-dog-like bright spot appearing just prior to sunrise and just above the point where the Sun will soon appear. It is rare, and differs from the normal sun dog in being vertically, not horizontally, displaced from the Sun. Also, it appears while the Sun is still below the horizon, unlike regular sun dogs that flank the visible Sun.

The false sunrise is caused by incident light from the (still hidden) Sun reflecting off crystals of ice in a layer of cirrus cloud. Flat crystals, if undisturbed, tend to orientate themselves parallel to the ground as they drift downward through our atmosphere and their flat face can act like a tiny mirror. Where these crystals exist in great numbers, the combined effect is like a single enormous mirror reflecting the image of the Sun. A similar phenomenon known as the *subsun* can sometimes be seen from aircraft flying over icy clouds. Here the image appears directly below the Sun as the natural composite mirror of ice crystals reflects sunlight back into space.

Another closely related spectacle is the sun pillar, a phenomenon caused by incident sunlight reflecting off the faces of plate-shaped or column-shaped ice crystals. This is distinguished from the false sunrise, both by being more common and, secondly, by assuming the form of a column of light either projecting above the horizon from a barely hidden Sun or flaring upward from the low Sun itself. Either way, sun pillars are not seen when the Sun is high in the sky, rarely occurring when the Sun is more than 6° above the horizon.

Incidentally, the Sun is not the only potential source of light pillars. The Moon sometimes is accompanied by a similar phenomenon and even artificial lighting (for example, street lamps) can cause pillars where abundant ice prisms are present at ground level. I am aware of an incident where moving columns of light were initially mistaken for a display of the aurora, only to be subsequently identified as light pillars caused by the headlights of motor vehicles reflecting from the myriad ice crystals filling the air on that very cold night!

Earth is not the only planet in the Solar System where sun dogs could be seen. Mars has both water ice and carbon dioxide ice ("dry ice") clouds, either of which is capable of producing sun dogs. Further out, crystals of methane and ammonia form clouds in the upper atmospheres of the giant planets and it is possible that truly spectacular solar haloes sporting as many as four sun dogs might be visible there – if only there was someone present to observe them. Maybe some day automated probes will send back images of these phenomena. Human beings are unlikely to directly witness them however, as the planets from Jupiter to Neptune are not exactly places that hold much welcome for astronauts!

"My Heart Leaps Up When I Behold a Rainbow in the Sky"

Better known than the sun dog is the *rainbow*. Indeed, it is surely only the frequency of this phenomenon that stops it from receiving its deserved adulation as one of the skies greatest spectacles. Rainbows can be truly beautiful, but we see them so frequently that even the best examples seldom receive more than a passing glance.

Fig. 6.3. Primary and secondary rainbows photographed in Sweden 14 July, 2009 (Credit: Jerry Magnum Porsbjer)

Like sun dogs, rainbows result from incident sunlight being split into its spectrum colors by myriads of tine naturally occurring prisms. But here the prisms are drops of water (falling rain), not crystals of ice. In the rainbow, incident sunlight follows a rather circuitous path. It is first refracted upon entering the raindrop, then internally reflected from the back of the drop and, finally, refracted a second time as it exists the raindrop. This has the effect of spreading the light over a wide range of angles, although the greatest intensity is confined to a narrow 40–42° range; the radius of a complete rainbow arc. This angle is not affected by the size of the drop, but does depend upon the refractive index of the water. This is constant for rainwater and for the fresh water spray of waterfalls and the like, but seawater has a somewhat higher refractive index and therefore rainbows formed by spray from the surf have a smaller radius than those seen in rain clouds.

The extent to which light is refracted depends upon its wavelength, perceived as its color. Blue light, which lies near the short wavelength end of the visual spectrum, is refracted at a greater

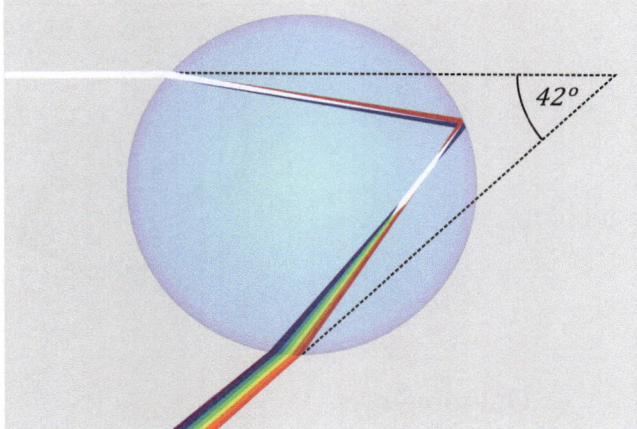

Fig. 6.4. Refraction and reflection of light in a raindrop. White light separates into different wavelengths (colors) on entering the drop. The different wavelengths exit the raindrop at different angles, producing a spectrum of colors, which we see as a rainbow (Credit: KES47, 4 June 2010)

Fig. 6.5. Rainbow in spray of the surf, Santa Cruz, California (Credit: Mila Zinkova)

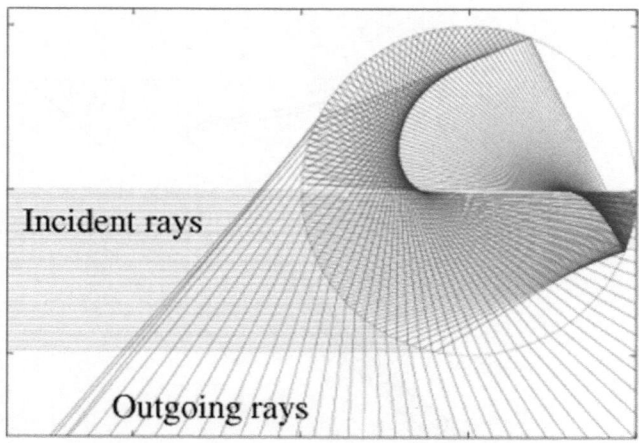

Fig. 6.6. Plot of rays in a secondary rainbow (Credit: CYD, English Wikipedia project, 6 June 2003)

angle than longer-wavelength red light. However, because it is also reflected from the back of the raindrop, it emerges from the drop at a *smaller* angle to the incident light than the emergent angle of the red. If the red light from a particular raindrop is seen by an observer, the blue light emerging from that same raindrop will be on a sufficiently different angle and will miss the observer altogether. The blue light which the observer actually sees is from a *different* raindrop; one that is below that from which the red light emerges!

Technically, no two people see the same rainbow. A rainbow is not really a "thing" out there in atmospheric space which we see, so much as interplay of refracted/reflected light and the eyes of diverse beholders. What is often said about beauty (wrongly, in the writer's opinion; but that is another story!) really *is* true of the rainbow. It is indeed in the eye of the beholder.

This is also true when one sees a rainbow reflected in the waters of a pond. At least, when one sees what *looks like* a rainbow reflected in a pond! The truth is, the rainbow seen by reflection is not the same as the one in the sky above it. Each results from different raindrops.

This situation permits the occurrence of a very strange phenomenon that is very rarely observed, and appears never to have been reported as far as the writer is aware. Maybe it has been seen but is so odd that those witnessing it are too afraid to mention it for fear of ridicule. ***The phenomenon is this; a reflection of a rainbow existing with no rainbow visible in the sky!***

Higher Order Rainbows

The sight of a second rainbow, fainter than the primary one and located "outside" of it, is not an unusual sight. These secondary bows result from a double refraction of sunlight within the raindrops and appear at an angle of 50–53°. This greater radius, in comparison with that of the primary, effectively diffuses the light of the rainbow to a greater degree than is experienced by the primary and, together with the greater leakage of light from two refractions, explains the fainter appearance of these secondaries.

The colors displayed by the secondary bow are inverted compared to those of the primary, giving it the appearance of a mirror image of the latter. The unusually dark region of sky between the two bows was first described by Alexander of Aphrodisias and is known as "Alexander's band" in his honor.

On rare occasions, a third (tertiary) rainbow has been reported and occasional sightings of quadruple ones are not unknown.

Fig. 6.7. Double rainbow at Wrangell-St. Elias National Park, Alaska (Credit: Eric Rolph)

In these latter instances, the "fourth" bow is said to have a pulsating or rippling look about its outermost arc.

Even more intriguing displays have been observed in which a secondary rainbow is seen to form unusually close to the primary at the same time as a more remote second rainbow disappears. During the transition, the remote "second" becomes a "third" or tertiary bow. The two remaining bows then merge into a band of white light, but with a blue tinge on the inner and a red on the outer edge. Before merging, the twin rainbows are clearly distinguished from the normal primary/secondary configuration, not only in their unusual proximity to one another, but by sharing identical color sequences – blue on the inside and red on the outside, just like a single or primary rainbow.

The rare quadruple or fourth-order rainbow appears to be as high as can be achieved in nature, although hypothetical rainbows up to the 19th order were theorized by Felix Billet (1808–1882) in the nineteenth century. Unknown, of course, to Billet, the development of the laser in the latter half of the following century has enabled the creation of higher-order bows in the laboratory. In 1979, K. Sassan produced the first sixth-order rainbow by hitting a pendant drop of water with a collimated light beam from a helium-neon laser. Then, in 1998, P. Ng, M. Tse and W. Lee used an argon ion laser beam to produce a rainbow up to the 200th order!

Supernumerary and Reflection Rainbows

On infrequent occasions, several faint rainbows are seen on the *inside* of a primary bow. Even more rarely, similar bows may be seen on the outside of a primary, but these *supernumerary* or *stacker* bows are clearly a different phenomenon to the more common secondaries.

These rainbows can only be explained in terms of interference between rays of light following paths having slightly different lengths within a raindrop. Because the phenomenon of interference can only work if light has a wave nature, the observation of supernumerary rainbows gave the first indication that light does indeed travel as a wave; a fact used by Thomas Young in explaining the phenomenon back in 1804.

Fig. 6.8. Supernumery rainbows 27 September, 2005 (© Andrew Dunn, reproduced under Creative Commons Attribution – Share Alike 2.0 Generic License)

Briefly, rays that are in phase reinforce each other by constructive interference and form a bright band, while those that are out of phase (by up to half a wavelength) cancel each other out through destructive interference and create a dark gap. Because light rays of different wavelengths (colors) have slightly different angles of refraction, these interference patterns differ slightly for different parts of the spectrum, causing the bright bands themselves to differ in color and form a small rainbow. Supernumerary rainbows are best seen when raindrops are uniformly small.

Earlier, we spoke about rainbows apparently reflected in bodies of water. These are sometimes simply referred to as *reflected rainbows* (not, as we said, a completely accurate description!) but this term is not to be confused with *reflection rainbow*, a confusingly similar expression used to denote an entirely different phenomenon.

A reflection rainbow is a one created by sunlight reflecting off the surface of a body of water. It is, one might say, a rainbow caused by a reflected image of the Sun rather than directly by the Sun itself. These rainbows are only visible if the body of water is large, quiet and close to the rain curtain. Because of these rather stringent

Fig. 6.9. Reflection rainbow over Skagit Bay, Whidby Is. WA. With the Sun setting and reflected behind camera from the waters of Dugualla Bay, 11 September 2007 (Credit: Terry L. Anderson)

requirements, they are not often seen, but when they are, they can be truly spectacular. Intersecting the primary rainbow at the horizon and reaching high into the sky, the centre of a reflection rainbow is as far above the horizon as that of the primary is below it. In (rare!) instances where a primary and secondary bow, together with their reflections, double with a reflection bow and *its* reflection, it is possible to see six or even eight bows simultaneously; a truly remarkable sight!

Titanic Rainbows?

Earth is not the only place in our Solar System where it rains. As we saw earlier, rain is also known to fall on Saturn's giant and exotic moon Titan. The biggest difference; whereas rain on Earth is in the form of drops of liquid water, on Titan it comes as oily drops of liquid methane.

Methane rainbows would be larger than water ones, thanks to the different refractive indices of these two liquids. Rainbows on Titan are titanic in more ways than one, with radii as large as 49° instead of the familiar 40–42° of their terrestrial counterparts.

There could be a problem however. Because of its great distance, the Sun as seen from Titan is relatively faint. Moreover, it struggles through an atmosphere smoggier than Los Angeles in the middle of summer. The faintness of the Sun in comparison with the view from Earth is not in itself a problem (although the rainbows of Titan would be less bright than the best of the ones we are blessed with) but the semi-opacity of the smog of organic particles that perpetually enshrouds the moon presents the chief difficulty. Titan's atmosphere is, however, more transparent to infrared wavelengths, so if a future astronaut to that distant world wishes to see a genuine extraterrestrial rainbow, a pair of IR goggles might be a sound investment!

Moonbows and Fogbows

Because raindrops are relatively large, a traditional – terrestrial – rainbow displays well differentiated bands of color. At night, the Moon can also produce rainbows, although they are far more rarely witnessed than the daytime solar ones. In these, the colors are a lot more muted. Bright *moonbows* can still be colored, albeit in more subtle hues, but fainter ones seem essentially whitish. Nevertheless, a whitish beam of light in the nocturnal sky is itself an impressive, if subtle, sight. Like daytime rainbows, moonbows can also occur in the spray of waterfalls, adding to the grandeur of an already majestic scene.

A *fogbow* is similarly pale in comparison with a rainbow, but in this instance the relative lack of color is due to the small size of the water droplets themselves, a mere 0.05 mm in diameter or smaller. At this size, diffraction tends to smear out the colors, rendering them less distinct than in a true rainbow. Where the fog droplets are very tiny, the colors are lost altogether resulting in a type of fogbow sometimes descriptively known as a *white rainbow*.

Fig. 6.10. Luna rainbow photographed from Zambian side of Victoria Falls, 2002 (Credit: Calvin Bradshaw)

The outer radius of a fogbow is also slightly smaller than that of a rainbow. In short, a fogbow appears as a sort of "watered down" version of the rainbow. Very pale ones can sometimes be produced by moonlight where they are known, naturally enough, as *lunar fogbows*.

Probably the most spectacular (and certainly the most startling) phenomenon associated with the fogbow is the so-called *glory* or *Brocken Specter*. This is most often encountered in mountain regions when a low Sun behind the climber projects his image onto a bank of cloud or fog. Suddenly, the unsuspecting climber is confronted by an enormous figure surrounded by a halo of pale rainbow colors! Can there be little wonder that mountains have long been held in superstitious awe by humans and deemed to be the haunt of giants and demigods?

A curious incident related to the writer appears to have involved a variation of the glory phenomenon. Driving home early one morning through heavy fog, a driver was troubled by another

Fig. 6.11. Fogbow and Brocken Specter 6 October 2007 (Credit: Mila Zinkova)

vehicle just a short distance ahead of him. If he (the driver) sped up, so did the other vehicle. If he slowed down, the other vehicle did likewise. This behavior started to worry the driver, as the road was almost deserted at that hour and one might seriously question the motives of anyone behaving in that way. After a little while though, the true nature of the "other" vehicle became clear. It was just an image of the driver's own reflected in the fog!

As I recall, this happened late at night, so I am not sure whether moonlight, or even the headlights of a genuine second vehicle following the witness' one, was the source of illumination. If the latter, the driver of the second vehicle must have been equally puzzled by the behavior of the witness!

Fingers of Light in the Twilight Glow!

Sometimes just after sunset or before sunrise, fingers of light are seen radiating outward, either from beyond the horizon at the place where the Sun has just disappeared or from where the morning Sun will shortly rise, or from above a low Sun shining amongst scattered clouds. The effect is quite common, but can be very

Fig. 6.12. Brilliant crepuscular rays over Santa Clarita, CA, 30 September, 2010 (Credit: Chevy III)

impressive. If the sky is clear and one faces the direction directly away from the sunset/sunrise point, what almost appears to be a reflection of the diverging rays will be seen on the opposite horizon. The rays there appear to converge opposite the Sun after having spanned the entire dome of the heavens. This apparent convergence is an effect of perspective, a la the oft-used example of apparently converging railway tracks with distance.

These shafts of sunlight are known as crepuscular rays; from *crepusculum* the Latin word for twilight. They are caused by shadows, either of distant mountains or of clouds (in both instances, often so remote as to lie beyond the local horizon) falling on the ever present particles of haze in the atmosphere. The rays themselves are really the regions of sunlit atmosphere that lie between these long shadow bands. Similar effects may be seen at any time of day when the Sun is hidden behind a bank of clouds or when a shaft of sunlight shines down through a hole in a cloudy layer. This latter is sometimes known as *Jacob's Ladder* or, when shining down on water, as *Sun drawing water*, reflecting a belief of the

Fig. 6.13. Lunar "crepuscular" rays. Though not strictly speaking "crepuscular" (that word refers to twilight, so to be pedantic, only twilight rays are truly "crepuscular") closely analogous phenomena can be displayed by the rising or setting Moon, as shown in this 2008 image (Credit: Mila Zinkova)

ancient Greeks that sunbeams draw water into the sky. These cannot truly be called *crepuscular* rays however, not being phenomena of the twilight.

Also not strictly describable as "crepuscular", similar rays may also be seen at moonrise or moonset when the Moon's face is sufficiently illuminated. These lunar "crepuscular" rays have the same causes as the solar ones; shadows of distant clouds or landforms breaking up the moonlight glow.

"Crepuscular Rays" in Outer Space?!

Crepuscular rays and their close relatives such as Jacob's Ladders are strictly phenomena of the atmosphere. They cannot occur in the vacuum of outer space or on worlds (for example, the Moon) where there is no atmosphere in any ordinary sense of the word. That is obvious, common sense, no possibility of being wrong! Crepuscular rays on worlds such as the Moon and Mercury simply cannot happen. Period!

Imagine the surprise, therefore, when astronauts on board *Apollos 8, 10, 12* and *17* reported seeing crepuscular rays extending up over the lunar limb! Imagine what the astronauts themselves must have thought when they started seeing something that by all accounts just could not have been happening. Fortunately, they all saw the same thing and many pairs of eyes could not be lying. Somehow, the impossible was happening!

Actually, these observations were not completely unprecedented, as the earlier *Surveyor* spacecraft also recorded horizon glows after lunar sunset and prior to sunrise. Yet, the drawings made by astronauts Gene Cernan and Harrison Schmidt from lunar orbit depicted phenomena so alike the familiar twilight rays of Earth as to stretch credibility had they not been so well observed.

Based on the experience of Earth, an "obvious" explanation presents itself. The Moon must have an atmosphere after all. But the "obvious" explanation is just as obviously wrong. The surface of our satellite sits in a hard vacuum. There are a few more atoms and ions near the surface than in empty space, but nothing capable of suspending particles and giving rise to twilight phenomena such as crepuscular rays. Clearly, the cause of these lunar rays differed from that of the terrestrial ones.

The clue to what was probably happening was uncovered by an instrument known as the *Lunar Ejecta and Meteorites* (LEM) detector left on the Moon's surface by the *Apollo 17* astronauts. Primarily intended to monitor the frequency of meteorite impacts

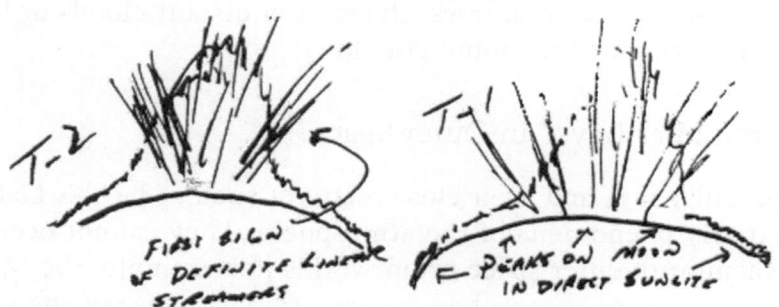

Fig. 6.14. Crepuscular-like rays on Moon as observed and sketched by Apollo 17 astronauts in 1972 (Credit: NASA)

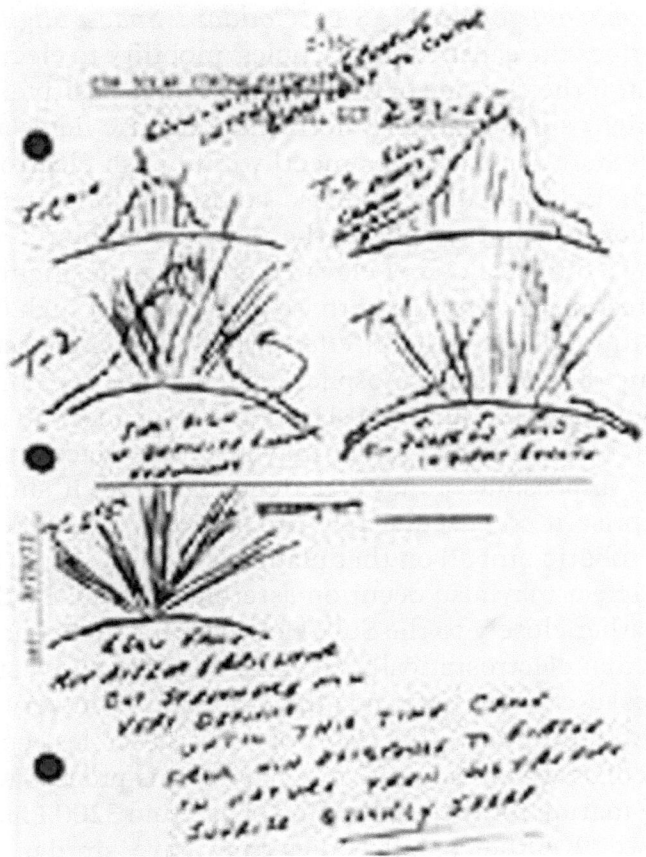

Fig. 6.15. Lunar sunrise phenomena as observed and sketched by Apollo 17 astronauts in 1972 (Credit: NASA)

by measuring the dust they kicked up as they hit the lunar surface, this instrument also detected a surprisingly large number of moving particles around the time of each lunar sunrise. There was no reason to suspect each sunrise to be heralded by a shower of meteor impacts and, in any case, the particles being detected were moving more slowly than those raised by impacts. Moreover, they were also highly directional, traveling from east to west across the terminator.

The most likely explanation for this strange discovery was the hypothesis put forward by Timothy Stubs of the *Solar System*

Exploration Division at NASA's *Goddard Space Flight Center.* Stubs ascribes the cause of the particles' mobility to electricity. He argues that if the day side of the Moon becomes positively charged and the night side negatively, horizontal electric fields at the terminator (where the two sides meet) would push electrostatically charged particles of dust sidewise across the terminator. Even without benefit of atmosphere, the terminator should always be marked by a diffuse cloud of elevated dust particles right along its length. Presumably, enough of these particles waft high enough to give rise to modified twilight effects, including crepuscular rays, that mimic the familiar atmospheric phenomena seen on Earth.

Although we (or our probes) have not as yet been in a position to observe it, something similar may also take place on Mercury. With the "impossible" lunar experience in mind, it should come as no surprise if crepuscular rays are someday reported (presumably by a robotic probe!) on that planet as well.

The effect may also occur on asteroids, especially those that venture rather closely to the Sun. However, for such small bodies, it may be that electrostatically elevated dust particles can (at least sometimes) escape the asteroid altogether and form very faint and attenuated tails of dust, like a sort of "dry" or ice-less comet.

Indeed, as we already saw in Chap. 1, this process was raised as an explanation for the brightening of asteroid 3200 Phaethon on June 20, 2009. Other small bodies may have similar outbursts through this same process, although observing such minor events will not, in general, be an easy task.

As an aside, we might note that bodies most vulnerable to this process would be the hypothetical "Vulcanoids" thought to orbit entirely within the region between the Sun and Mercury. Remnants from the original solar nebula and later editions in the form of fragments blown off the surface of Mercury by violent impacts should inhabit the Vulcanoid zone, but as of mid-2011, not one has been discovered. If any Vulcanoids exist, they must be very small. Is it possible that they have been largely eroded away by the electrostatic effect? It would be interesting to see just how significant this process of erosion may have been on asteroids confined to Sun-hugging orbits throughout the lifetime of the Solar System, especially if searches for Vulcanoids continue to turn up negative results.

Dark Days and Atmospheric Opacity

Believe me, one of the most unnerving experiences that we can have is to wake up to absolutely no light. Not even the faintest glimmer! Not long ago, this writer had just such an experience. Waking in the early morning hours, there was simply no light at all. I was on the western plains of New South Wales at the time, so the glare of streetlight and so forth was not expected, but even a cloudy night in a rural setting has some light; at least, that had always hitherto been my experience. And on a clear night, the combined illumination of a starry sky and airglow (to say nothing of even a slender Moon if it is present) throws a surprising amount of light onto the land. But on this night – nothing. For a few moments, I began wondering if I had gone blind in my sleep, and only a faithful flashlight next to my bed fully dismissed such a dismal thought.

From my location, the next day dawned relatively clear. Not so for Sydney and the coastal regions. There, residents awoke to a red glow suffusing everything. So strange was the light that some wag at one of the radio stations thought the song *End of the World* would be a suitable disk to spin.

It was not the end of the world of course, but it probably signaled the end of career for many country farmers. What had caused the deep darkness at my location and the weird redness at Sydney was a huge cloud of dust; topsoil blown from drought ravaged farms by strong winds and lofted far and wide by the prevailing westerly stream.

Dust storms like this, as well as smoke from large forest or prairie fires and volcanic eruptions have been responsible for many instances of dark days and strangely colored daylight. In times past (and not only in times past!) these have caused much fear. This is entirely understandable, as the apparent failure of sunlight speaks to some of our deepest fears. Yet, if these events are explained relatively simply by atmospheric opacity resulting from dust, smoke or volcanic ash, there is no real anomaly, no real mystery, here. We should be able to look back on records of dark days and find everything according with this explanation. The trouble is, we do not always find this. Many dark days fit perfectly, but some do not.

What, for example, can we make of the account from April 23–25, 1547 which the famous astronomer Kepler relates on the authority of Gemma that "the Sun appeared as though suffused with blood, and many stars were visible at noonday"? This was clearly not a solar eclipse, but if the dullness of the Sun and apparent darkness was caused by atmospheric opacity, how were stars visible at noonday?

That great compiler of anomalous reports, William Corliss, wondered why, if all dark days and similar events result from dust, volcanic ash and fires, there should be an apparent lack of any reports of these types of events coincident with some of the dark day reports. Where, for example, is there evidence of fires or the like at the time of the Washington dark day of September 12, 1902? Why is there seldom a mention of the smell of smoke or fall of ash in the accounts of many dark days? Certainly, some dark days can be traced to these events, but the causes of others are more doubtful. Some people have even speculated that at least a few might have resulted from the encounter between Earth and clouds of cosmic dust. This sounds a bit farfetched, but the universe that we inhabit is such a strange place that it cannot simply be dismissed on that ground alone.

In fact, there is one recorded instance of, if not a "dark day" in the full sense of the term, at least a day of odd sky effects and apparent atmospheric opacity of a milder degree, that almost certainly had an astronomical origin. This happened on June 30, 1861; a day on which peculiar sky effects were reported from widely separated regions of the planet.

In England, "Mr. Lowe of Highfield House" noted that the sky took on an odd yellowish tint late that afternoon, at which time the Sun also appeared unusually dull. Although stopping well short of a fully fledged dark day, there must have been a noticeable diminution of daylight as Lowe mentioned that the local vicar had the pulpit candles lit at 7 P.M., still well before sunset on that long summer's day. Similar sky effects were also noted by others, including the well known astronomer J. R. Hind.

These sky anomalies did not stop at evening. Neither were they confined to the northern hemisphere. On the other side of the world, in New South Wales, astronomer John Tebbutt noted that the eastern horizon was illuminated that evening, as if the Moon were about to rise, although the Moon was then past last quarter and rising hours later than the time of this observation. Various reports of "aurora" (?) were also received from wide areas of the state at that time.

The odd daylight sky, apparent dimness of the Sun and nocturnal illuminations of that date almost certainly had an astronomical source. The culprit was a large comet discovered on May 13 by the same John Tebbutt who observed the luminous eastern sky of New South Wales. Initially a telescopic object, the comet passed close to Earth at the end of June and became bright enough to be visible in full daylight without the use of optical aid. On June 30 – coinciding with the strange sky phenomena – the Earth passed through its dusty tail. Now, we cannot prove with absolute certainty that this encounter with the tail gave rise to the nocturnal glows, sensation of dullness and so forth reported on that day, but it would be an enormous coincidence if the two were unrelated.

Which raises an interesting question. Were some other sky peculiarities – even genuine dark days – similarly caused by encounters with cometary debris?

The one which primarily stands out in this respect is an event recorded by Erman for February 12, 1106. According to this account, the Sun was darkened on that day and "meteors" appeared in the sky. On the face of it, the reference to "meteors" is especially suggestive of cometary debris, although the term was sometimes used to cover a broader range of atmospheric phenomena than its stricter usage of today.

In any case, at the time of this event, a bright comet with a very long tail was visible in the skies. Earlier that month, records show it to have been visible in the daytime close to the Sun – shades of Tebbutt's Comet of 1861!

Before we can jump to any conclusions however, it is worth mentioning that, although no satisfactory orbit has been computed for this comet, it is widely believed to have been a member of the so-called Kreutz group of sungrazers. These comets pursue similar orbits that take them (as their name implies) very near the Sun. Some smaller ones have even hit it! But (and here's the problem) if the 1106 comet truly was a Kreutz object, it could never have approached closely to Earth and the Earth/comet/Sun geometry would have precluded any possibility of Earth's passing through its tail or, for that matter, encountering any other form of detritus from the comet. The daylight visibility, unlike the 1861 object, resulted from the comet's very close encounter with the Sun, not with the Earth.

If the majority of astronomers are right about the 1106 comet, the dark event of February 12 must have had some other, unrelated, cause. Of course, it *is* possible for astronomers to be wrong sometimes, so it might be suggested that the comet was not a Kreutz sungrazer after all, that it did make a very close approach to our planet and that we did pass through the tail (or a cloud of debris) on February 12. The biggest difficulties with this, apart from the good agreement between the comet's position and appearance and that of a large sungrazer appearing at that time of year, is the gap of over a week between the comet's daylight showing and the dark day (unlike Tebbutt, where the two coincided as the comet passed almost between Earth and Sun) and the fact that the comet remained in the western evening sky through the date of the dark day. Because of the geometry necessary for an encounter between Earth and a comet's tail, one would normally expect a comet to go from morning to evening sky (or vice versa) or from northern to southern hemispheres (once again, or vice versa) as the comet passed between Earth and Sun. In the 1106 instance, the comet simply became more prominent in the western sky as it drew away from the Sun. Had the dark event happened earlier or the daylight sighting later, there may have been a case to answer. As it is, the comet appears almost certainly innocent of causing the darkness recorded that year.

Dry Fogs

From time to time, the lower atmosphere becomes strangely opaque as if enveloped by a sort of fog. Nevertheless, the reports indicate that this is no normal water-vapor fog, but something more closely resembling dust or solid particles. Moreover, these "dry fogs" can apparently cover very large regions of the Earth's surface. Although not directly associated with dark days, they seem to be part of the same general phenomenon.

Like dark days, many so-called dry fogs undoubtedly result from dust or smoke being wafted into the air from large scale dust storms or forest fires. Volcanic eruptions are also notorious for covering large regions in fine dust and ash particles in what could certainly be described as a "dry fog." Yet, as with dark days, certain instances stretch such common sense explanations to the breaking point.

At the more extreme end of the "explicable" category, we may cite the long lasting "fog" of 1914 that occurred in the Sumatra region. An account of this phenomenon was published the following year in the pages of the prestigious science journal *Nature*. The "fog" appears to have been of remarkable duration apparently lasted throughout the months of September and October, disappearing only "when the rains fell" at the onset of the wet season. Writing about this phenomenon, C. Braak noted that, day after day "the Sun was not visible before 8 o'clock, when it appeared as a dim red disc. Objects in general were barely visible at 500 m distance, and on the thickest days invisible within half this distance." This "fog" appears to have been an unusually thick and persistent dust haze, probably resulting from dust raised by desert winds and spread throughout the lower atmosphere over a wide area. It seems that the affected area quite frequently experiences similar events; another major one that occurred in 1902 having already been noted in the journal *Science* 4 years later.

Far less readily explained is the strange "fog" that was said to have enveloped much of the planet during the summer of 1783. So strange are the tales of this event, that we may be tempted to dismiss the whole episode as legend, were it not for the relatively frequent accounts of the phenomenon in scientific articles during the following century and beyond.

The "fog" (if such it can be called) is said to have begun around May 29 that year and to have extended "from Africa to Sweden, and throughout North and South America" (!) It persisted, according to the reports, for over 1 month, seemed unaffected by wind direction and was not dispersed by rainfall. The Sun allegedly became visible only when it attained an elevation of around 12°, and to have continued faint and red throughout the rest of the day. A "disagreeable smell" was also noted in several places. But most unusual were reports that, during the night time hours, the fog appeared to be self luminous and "afforded even at midnight a light equal to that of the full moon, and which was sufficient to enable a person to see objects distinctly at a distance of 200 yards."

Some speculated that this global fog might have marked the passage of Earth through the tail of a comet. The duration of the event, not to mention the lack of any suitable comet visible at the time, makes this explanation very dubious to say the least. Moreover, it would need to be quite a comet tail to cause the reported effects. The writer would certainly like to see its equal!

A more reasonable suggestion holds the eruption of Mt. Hecla responsible. If so, a tremendous amount of fine dust must have been lofted high into the atmosphere, while a substantial quantity must also have been restricted to ground level if reports of an odor are to be believed. Dust at very high altitudes may have reflected sunlight to cause the bright nights (the occurrence of the event during late spring/early summer season goes some way in supporting this), but whether all of the phenomena associated with this fog can be explained as volcanic in origin remains uncertain.

Incidentally, a similar worldwide "fog", likewise accompanied by bright nights, is said to have occurred in 1831, but details of this event are sketchy. It is also interesting to compare the bright nights and haze reported over much of Europe following the great Tunguska meteor explosion of 1908 with these earlier accounts. Although there is no direct reason for thinking that Earth encountered giant meteorites in 1783 or 1831, the possibility remains an interesting one to speculate upon!

Blue Moons, Blood Suns and Screaming Artists!

The expression "blue Moon", synonymous with rarity, has two distinct meanings. The one which probably gave birth to the phrase "once in a blue Moon" most likely refers to the popular name given to the second full Moon in any calendar month. This happens, but not very often, hence the use of the expression. Needles to say, the second full Moon in a month is no more likely to be colored blue than any other!

In its second sense, the expression is taken literally. The Moon really does appear blue. This happens even more rarely than a month's second full Moon, but it does happen!

Indeed, not only blue Moons but also blue Suns are seen from time to time. These effects, although they can look strange and may even cause alarm, are merely the result of very small particles of nearly uniform size drifting through the atmosphere at altitudes of around 5 miles (roughly 8 km). Depending upon the exact size of the particles, certain wavelengths of light are absorbed while others are permitted to pass through, and when the red end of the spectrum of sunlight is absorbed, the Sun (or the Moon) will take on a bluish or even greenish hue. Fine smoke particles from forest fires can produce this effect and the smoke palls from major conflagrations have been known to produce blue Suns/Moons for several days in a row even hundreds of miles from the source of the fire. More commonly, the opposite effect is witnessed at times when the Sun or Moon hang low in the sky and their light passes through a maximum depth of atmosphere charged with dust or smoke. Then, light at the blue end of the spectrum is absorbed and only the red transmitted. On such occasions as these, the Sun or Moon can best be described – in graphic if slightly macabre phraseology – as looking like a ball of blood.

Volcanic dust, capable of shrouding much of the planet following a very major eruption such as that of Krakatoa in 1883, can turn not just the Sun but the entire sunset sky into the color of blood. Some have speculated that the lurid sunset skies may have found their way into art as inspiration for the background of *The Scream;* the famous painting by Norwegian expressionist painter Edvard Munch. The strange and tortured mummy-like figure of the screaming subject appears in a scene bathed in the light of a spectacular but strangely menacing sunset. Multiple colored clouds blaze like tongues of fire in the sky, casting an almost apocalyptic sense of foreboding over the scene. As this work was executed in 1893, just 10 years after the eruption of Krakatoa the question arises as to whether this was a creation of the mind of the artist – the lurid sunset expressing the mood of the subject – or whether there was a degree of realism about it. Was this scene based on an actual sunset; a sky turned into blood by the volcanic dust of Krakatoa?

Fig. 6.16. "The Scream" (1893) by Edvard Munch (1863–1944). The National Gallery, Onslo, Norway

On the other hand an entry in Munch's diary headed *Nice 22. 01. 1892* suggests that the inspiration may have been more recent than a memory from 1883 to 1884 when the volcanic sunsets were most noticeable. Munch writes;

> I was walking along a path with two friends – the sun was setting – suddenly the sky turned blood red – I paused, feeling exhausted, and leaned on the fence – there was blood and tongues of fire above the blue-black fjord and the city – my friends walked on, and I stood there trembling with anxiety – and I sensed an infinite scream passing through nature.

Perhaps what the artist saw was nothing more than high cirrus cloud lit by the rays of a recently set Sun; the somber interpretation of the scene saying more about his mood at the time than the sunset's appearance per se. But may it also have brought to life memories of the strange sunsets of 8 or 9 years earlier? We will probably never know for certain, and with these few words, we should probably leave the possible Krakatoa connection in the capable hands of Munch scholars.

The Thing That Is Not There!

How many times has the tale been told? A traveler lost in the desert sees a lake in the distance and sets out toward its life-saving waters. But no matter how far he travels in the direction of the lake, it never seems to get any closer. It is as if the water keeps drawing away from him, a little like Homer's Tantalus whose punishment in Hell was to become almost totally immersed in water only to have it swallowed by the ground the moment he bent forward his head to drink and assuage his extreme thirst. The desert lake is equally tantalizing. The moment the desperate traveler takes one step toward it, it seems to retreat by the distance of an equal step. Tragically, the traveler dies of thirst without ever reaching water.

Alas, such stories are not confined to fiction.

We all know what has happened of course. The hapless traveler was led astray by that trickster of the desert, a mirage.

Although mirages are widely thought of as desert phenomena, these hot and dry regions are by no means the only places where they occur. However, before saying anything further about where they may be seen, a common misconception must first be

Fig. 6.17. Desert mirage at Primm, Nevada, 4 April 2007 (Credit: Mila Zinkova)

cleared up. Mirages are *not* (despite what some dictionaries say) optical illusions, strictly so called. They *can* be photographed and filmed. They are "real" or objective phenomena existing in the world, not confined to the human eye or mind. Mirages are "real" in the sense that, say, a reflection in a mirror is real and equally capable of being photographed.

A mirage is essentially an atmospheric optical phenomenon in which light rays are bent to produce a displaced or in some way distorted image of either a terrestrial or astronomical object.

There are two broad types of mirages:

(a) Inferior mirages. These occur over land or water surfaces that are warmer than the air at slightly higher altitudes. The air immediately in contact with the hot surface is itself heated, expands, and is reduced in density. Light rays entering this layer of warm, low density, air are refracted upwards, much as they would be reflected upward upon striking a body of water. An observer of these upward refracted rays therefore sees the object from which they have been emitted (or reflected, which ever the case may be) as if it were reflected in a pool of water. Light rays traveling from the object through the cooler and denser air above the warm layer travel directly to the eye of the observer and give an accurate image, but those bending upwards into the observer's eyes produce a second image which looks exactly like a reflection of the first in a body of water. This is the traditional desert mirage.

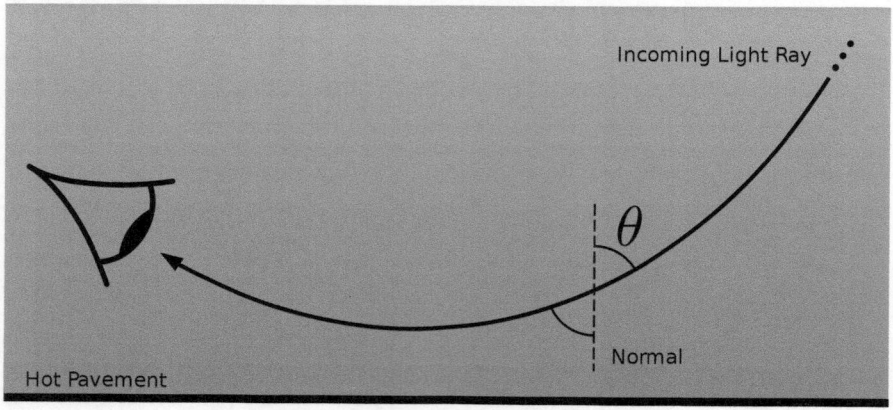

Fig. 6.18. Diagram demonstrating an inferior mirage over hot pavement (Credit: Rick Manning)

It is also very common on bitumen roads where the road takes on a striking appearance of being covered by water in which distant street signs and oncoming vehicles appear "reflected."

(b) Superior mirages. Normally air temperature decreases with altitude, however, under certain conditions, a layer of air close to or in contact with the ground can be colder than the air lying immediately above it. For instance, on a day of hot off-shore breezes coupled with a cold ocean current, the air in contact with the ocean surface will be colder than that a little higher up in the atmosphere. Similarly, on a calm day, the air immediately above snow or ice covered ground may be quite a deal colder than that at slightly greater height. In these circumstances, the reverse of the conditions favoring inferior mirages occurs, with the colder and denser air lying beneath the warmer and less dense layer. Light rays are refracted downward as they pass through the denser air, tending to follow the slope of the ground and the curvature of the Earth's surface, making it possible for an observer within the cold layer to see objects that would normally be out of his line of sight or even beyond the horizon. These mirages are known as superior mirages, and are frequently seen on the polar ice fields, although they may occur anywhere the above conditions are met. The writer has seen several over the ocean on hot days, but only while standing very close to sea level. A slight rise in altitude puts the observer above the cool air layer and the mirages just seems to sink into the ocean.

Remarkable examples of this form of mirage have been reported by Antarctic explorers. Far out on the ice, hundreds of miles from open water, ships have been seen sailing upside down on the horizon, interspersed amongst the clouds. On one occasion, an explorer's camp about 2 miles distant appeared to swell into a city of pyramidal skyscrapers before vanishing completely after nothing more than a shift of wind direction. The observer of this event dropped to his knees upon witnessing this (we are not told why!), and upon so doing, had the "city" reappear before him. Standing again to his full height, the mirage once more disappeared. Evidently the layer of cold air responsible for refracting the light reflected off the camp tents (at least, what remained of the layer after the wind change had apparently stripped the top from it) terminated somewhere between the observer's waist and his head!

Fig. 6.19. Superior mirage of Point Reyes as seen from San Francisco (Credit: Mila Zinkova)

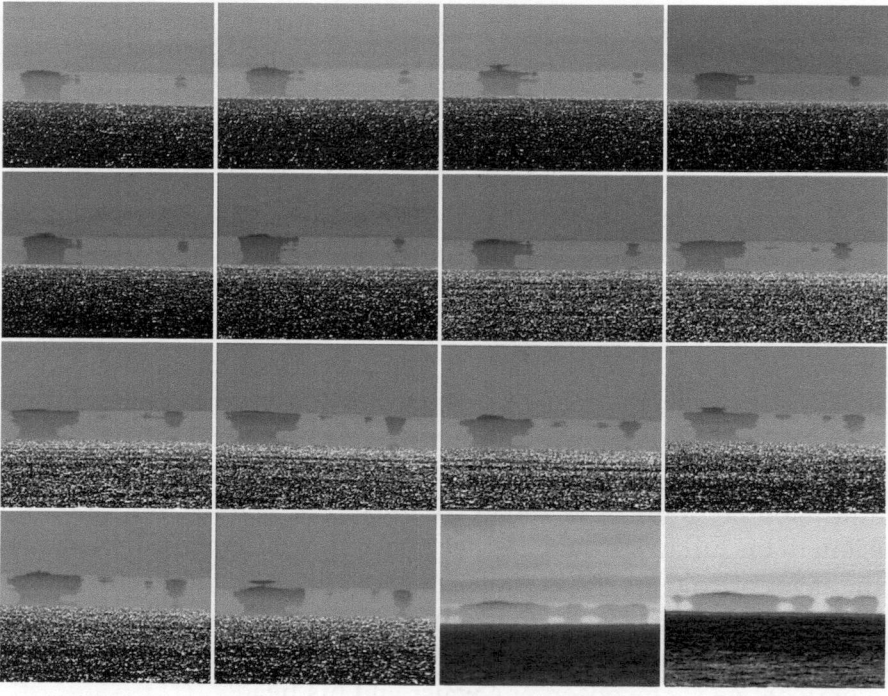

Fig. 6.20. Fata Morgana mirage of Farallons Islands (Credit: Mila Zinkova)

The example of the phantom ships seen hundreds of miles from water is an instance of a special type of superior mirage known as *Fata Morgana*. These mirages occur when the thermal inversion is so strong that the refraction of light rays exceeds the curvature of the Earth's surface and objects beyond the horizon are thereby made visible. Fata Morgana mirages can occur in both hot and cold conditions (in hot conditions, where the surface is colder than the air) but are most common near the poles. The true Fata Morgana is more complex than ordinary superior mirages and is made up of several images of the "miraged" object … upside down, right way up and stacked one upon the other! In calm conditions, atmospheric ducts – which may be thought of as naturally occurring refracting lenses – form where layers of warmer air rest on top of layers of colder, producing a series of both erect and inverted images. Needless to say, the object can be tremendously distorted and can end up looking like some other thing altogether (instanced by the camp becoming a city of skyscraper tents, although that mirage was not strictly a Fata Morgana).

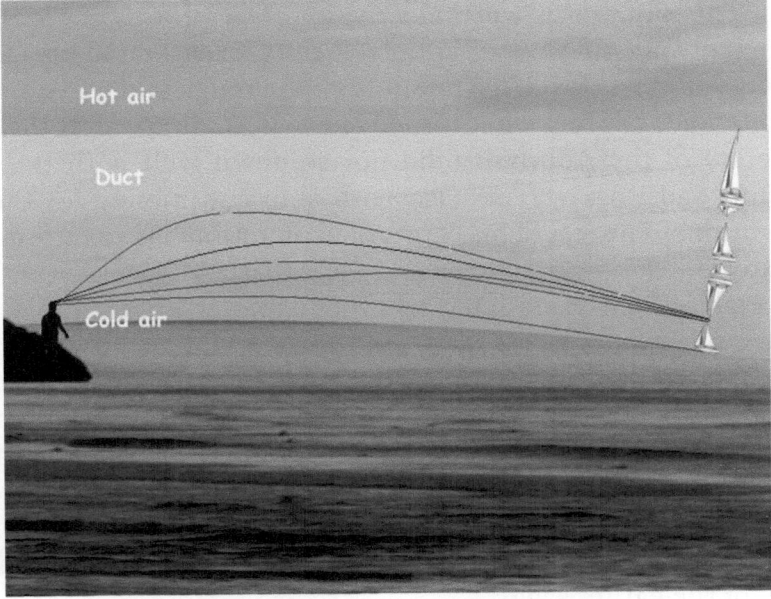

Fig. 6.21. Schematic diagram of a Fata Morgana mirage (Credit: Mila Zinkova)

Fata Morgana mirages are not confined to the daylight hours. As we will see in the next chapter, under suitable conditions, light rays from a source such as the headlamp of a motor vehicle can be bent to such an extent as to follow the Earth's curvature over vast distances and appear as "spook lights" whose distance from the observer can be very deceptive. As we shall see, at least some instances of this phenomenon in central Australia have contributed to the legend of the Min Min Light, a strange "ghost light" said to haunt the Australian "outback" and called by the native Koori people the "dead man's camp fire." But more of this anon!

Phantom Mountains and Lands of Mist

In the year 1818, while in search of the elusive Northwest Passage, Sir John Ross found his search blocked by what he took to be a land mass topped by a range of high mountains. Naming the range "Crocker Mountains" in honor of the First Secretary of the Admiralty, he abandoned his search for the Passage and returned home to England, despite the protests of some of his officers. One of these officers, William Edward Parry, continued the journey the following year without any encounter with the Crocker Mountains. The mountains were, clearly, a mirage of the Fata Morgana type. Probably they were simply greatly exaggerated images of icebergs or maybe even ice mounds on the rough surface of sea ice. In any event, naming what turned out to be a mirage after the First Secretary of the Admiralty did not go down well with the high officials and Ross was refused both ship and money for further expeditions. The rest of his career of exploration had to be financed privately.

Eighty eight years later, in 1906, explorer Robert Peary, saw what he believed to be a large land mass to the northwest of the highest point of Cape Thomas Hubbard. The estimated position of the land was given by Peary as north 83°, west 100°. By a strange co-incidence, he named the land Crocker Land, *not* after the same Crocker as Ross sought in vain to honor, but after Peary's late colleague George Crocker. The identity of name was a pure coincidence, as indeed, was the similarity in the name *Peary* and the *Parry* of the earlier saga.

Seven years later, Donald MacMillan led an expedition to reach and explore Crocker Land. On April 21 of the year 1913, the expedition caught its first sight of Peary's discovery, described by Mac-Millan as a land of "hills, valleys, snow-capped peaks extending through at least 120° of the horizon." Quite an extensive place it would seem, and no doubt the Captain and his colleagues were excited by what they were seeing. Nevertheless, the expedition had one resident skeptic. Piugaattoq, an Inuit hunter with 20 years experience in the region, insisted that what they were looking at was not a real land at all; merely a "poo-jok." This term apparently meant something like "mist" in the Inuit language. Though not a "mist" in the strict sense, apparently what he implied was something equally insubstantial; in other words, a mirage. Crocker Land and Crocker Mountains may not have shared the same Crocker, but they apparently shared the same nature. Neither was real!

MacMillan was not easily convinced however, and for 6 more days pressed on toward the elusive Crocker Land. After covering over 100 miles through dangerous sea-ice and finding Crocker Land as far away as ever, he reluctantly conceded that Piugaattoq was right after all and (red faced, as we may assume) abandoned the expedition.

The Silent City of Alaska

A story not unlike the Crocker incidents is told by early prospectors and native residents of Alaska. According to these witnesses, a city sometimes seems to materialize over a glacier in the Mount St. Elias region. The "silent city" is said to be so clearly visible, that houses, streets, cathedral spires and trees can be readily discerned.

In the summer of 1897, the duke of Abruzzi led an expedition in search of the city. He was not to be disappointed. As C. W. Thornton, one of the expedition's members wrote, "it required no effort of the imagination to liken it to a city, but was so distinct that it required, instead, faith to believe that it was not in reality a city." Some folk suggested that it was a mirage of the city of Bristol, 2,500 miles away in England – on the other side of the North Pole!

More likely, the mirage (which was reported regularly between June 21 and July 10 each year around the turn of the twentieth century) was a Fata Morgana of various objects at closer range (though still beyond the local horizon) whose images were distorted into the appearance of houses, trees etc. We are reminded as to how the Antarctic camp was transformed into a city of skyscraper pyramids. The silent city of Alaska was probably a somewhat more complex counterpart of this.

The Mirage of the Rising Sun

The ill-fated north polar expedition led by Willem Barentsz in 1596/7 left an unexpected legacy in the form of the discovery of a curious mirage phenomenon. Whilst navigating the island of Novaya Zemlya off the northern coast of Siberia, one of the members of the expedition – Gerrit de Veer – noted that an image of the Sun could at times be seen above the horizon, even though the real Sun was still up to 5° below it. Toward the end of the long polar night, the Sun is not altogether absent from the sky! Often, however, its appearance is vastly distorted as the following description by Fridtjof Nansen describes;

> The mirage was at first like a flattened-out glowing red streak of fire on the horizon; later there were two streaks, one above the other, with a dark space between; and from the main-top I could see four, or even five, such horizontal lines directly over one another, and all of equal length; as if one could only imagine a square dull-red sun with horizontal dark streaks across it.

Nansen noted that at the time of these observations, the Sun was over 2° below the horizon and would not climb above it for several more days.

This type of mirage phenomenon has been named the *Novaya Zemlya effect* in honor of the island from where it was first specifically recorded, thereby giving this remote and frigid land mass at least one claim to fame. The N-Z effect is not, however, restricted to Novaya Zemlya, nor even to the frigid lands of long nights. Indeed, it may occur where ever temperature variations are sufficiently large to cause a high degree of refraction. Nor is it only the Sun that can experience this effect. The Moon and other bright astronomical objects are not immune and (as we shall see in the following chapter) Novaya Zemlya images of bright planets such as Venus might explain some spook light reports such as those of Central Australia.

Whilst we are on the subject of the rising and setting Sun, mention should be made of a beautiful phenomenon occasionally seen when the Sun (or sometimes the Moon or – occasionally – Venus) is observed to set (less often, to rise) over a level horizon such as the ocean, an expansive flat plain or an ice field. Because the depth of atmosphere through which the object's light passes acts somewhat like a prism, different colors are separated, with the red end of the spectrum being more readily absorbed by the depth of air through which it travels. By contrast, the air is more transparent to green light and under conditions of clear and clean air, the last portion of the limb to set – or the first part to rise – can turn a spectacular green. If the surface air is warmer than that immediately above it, an inferior mirage of this *green flash* can (in the words of one eyewitness) "[flood] the desert like an alien helio-graph." When one deems conditions to be suitable, it is worth while keeping an eye on the setting Sun!

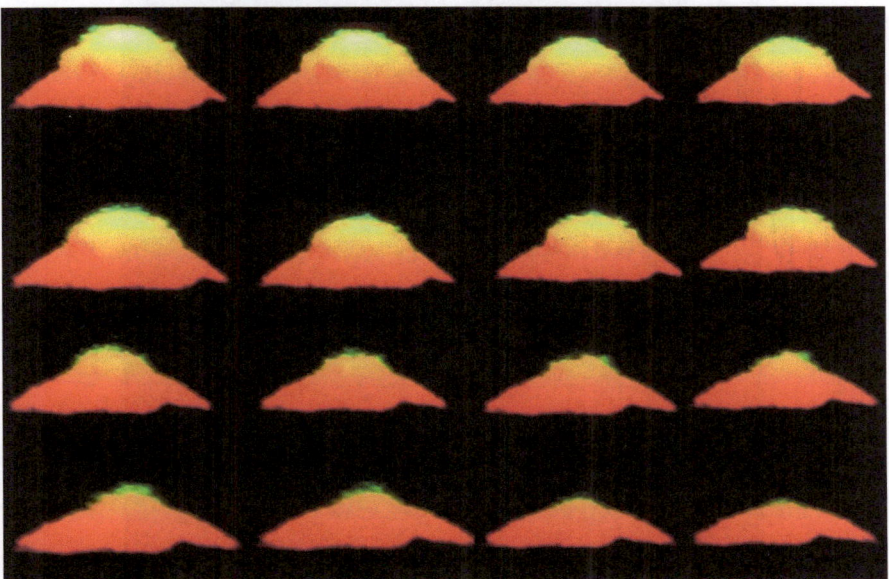

Fig. 6.22. Green rim & green flash of setting Sun (Credit: Mila Zinkova)

When several inversion layers exist – cold air trapped beneath warmer air which is itself overlain by another layer of cold air – a so-called *mock mirage* of the setting Sun or Moon can produce even stranger effects. On these occasions, the setting Sun or Moon can appear to break up into pancake-like layers separated by dark intervals. Although this appearance mimics Nansen's description of the Novaya Zemlya mirage, the similarity is one of appearance only. A mock mirage of this type is not an instance of the Novaya Zemlya effect, merely a look-alike.

Early in 2007, an inferior mirage of a setting comet was observed. This is an exceptionally rare phenomenon, as the comet involved must be unusually bright and of high intensity for such a thing to be observable. Alas, comets like that occur all too infrequently, but the one in January 2007 – formally known as C/2006 P1 McNaught – was the most brilliant for nearly 42 years and even became visible in full daylight for a short time. As it set in bright

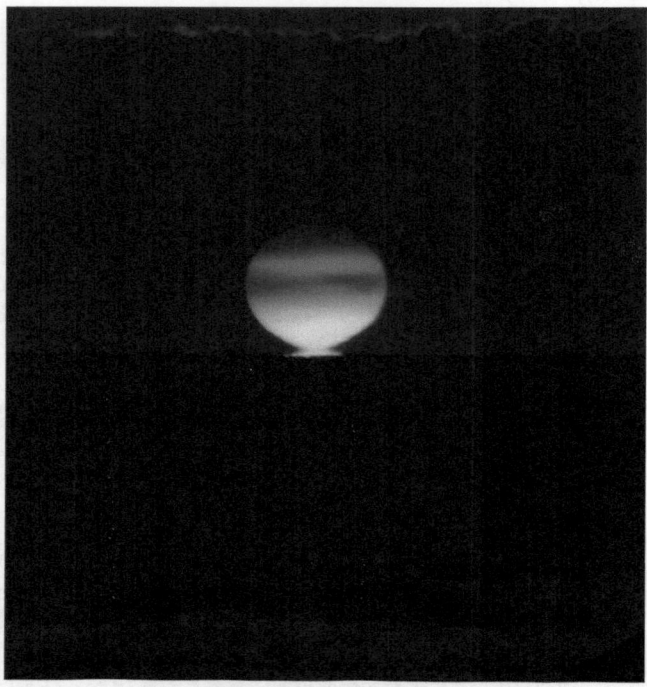

Fig. 6.23. Inferior mirage of setting Sun – the "Etruscan Vase" stage (Credit: Mila Zinkova)

Fig. 6.24. The "Omega stage" of setting Sun mirage 12 January 2007 (Credit: Mila Zincova)

Fig. 6.25. Complex mock mirage of sunset in San Francisco. The "pancake" phase of the phenomenon mimics some classic descriptions of the Novaya Zemlya effect, but the processes involved are quite different (Credit: Mila Zinkova)

Fig. 6.26. Inferior mirage of the setting Comet 2006 P1 McNaught, 12 January 2007 (Credit: Mila Zinkova [Mbz1 at en.wikipedia])

twilight, at least one fortunate observer was able to observe an inverted image of the comet seemingly rise up from the horizon to meet it; a most unusual sight!

A Brush with Polarized Light

The sense of color is a wonderful gift that we humans possess. Although we could survive quite well without it, just imagine how dull the world would be in black, white and shades of grey! The visual arts may not have progressed beyond the strictly utilitarian images of the most primitive cave art; those images showing which animals are good for food and which are to be avoided.

Yet, for all our abilities at seeing the shades of the visual spectrum, there are further "colors" which largely elude us, but which are readily perceived by some other species with which we share this planet. And I am not talking about electromagnetic wavelengths a little longer or shorter than visible light, although the same could indeed be said of these. The extra "colors" that I mean refer to the different visual sensations created by polarized light as perceived by such creatures as insects, cephalopods, fish etc. Not that "color" is really the right word in this context, but it is probably the best we can call upon under such circumstances as the experience of clearly perceiving the polarization of light is so far beyond our ken that we have no word to describe it. After all, how can we know from the "inside" as it were, how a fish or a honey bee sees the world?

Note however that we said "clearly perceiving." Although the fact is not widely appreciated, we human beings are not completely blind to the polarized property of light. We can, with practice, visually determine whether light is or is not strongly polarized as well as distinguish between linearly polarized and circularly polarized light. Moreover, if the light we see is circularly polarized, we can even tell whether the polarization is left or right handed. We can also directly "see" the polarization of the Sun's light, especially at angles of 90° from the Sun itself, where polarization is highest. It is as if we have a little used and little known extra visual sense, but what this sense perceives is not so much an extra color as a distinctive visual pattern.

We will look at this pattern shortly. First though, what exactly do we mean by "polarized" light?

Ordinary, non-polarized, light is comprised of electric and magnetic vibrations that take place in all possible planes containing the ray. These vibrations are perpendicular to the direction of the path of the ray and are themselves perpendicular to one another. That is to say, for each electric vibration, the associated magnetic vibration takes place in a plane at right angles to it. In plane-polarized or linearly-polarized light, the electric vibrations are confined to only one plane, known (not surprisingly) as the plane of vibration. Consequently, this also confines the magnetic vibrations to a single plane, namely that perpendicular to the plane of vibration. This plane is known as the plane of polarization. Circularly polarized light is defined as light which can be resolved into two vibrations lying in planes at right angles, of equal amplitude and frequency, but differing in phase by 90°. The tip of the electric field vector, at a fixed point in space, describes a circle around the direction in which the light ray is propagating. The direction of circular polarization may be either left handed or right handed.

So how may we "see" the polarization of light?

The first hint that we have this ability came in 1864 thanks to Austrian mineralogist and geologist Wilhelm K. von Haidinger (1795–1871). As Haidinger studied minerals under polarized light, in an effort to discern some tell-tale pattern in the refracted light, he noticed the appearance of a faint yellowish brush-like "stain" which persisted while ever he was looking directly toward the source of the polarized light. That is to say, it remained even when a mineral crystal was taken away from in front of the light-source. This proved to Haidinger that its origin must lie in the polarized light itself and not in refraction through the crystal. Moreover, as he rotated the polarizer, the "stain" rotated with it, adding extra proof that he was truly (and unexpectedly!) perceiving the polarization itself.

The "stain" is now known as *Haidinger's brush* in his honor. Observers mostly describe it as having the appearance of a diffuse and elongated yellowish pattern, with a pinch at the center giving it a somewhat hourglass look. Bluish leaves, generally shorter than the yellow formation, emerge from the central regions of the brush, perpendicular to its long axis. The brush is not large, occupying some 3–5° of the visual field or about 6–10 times the apparent diameter of the Moon.

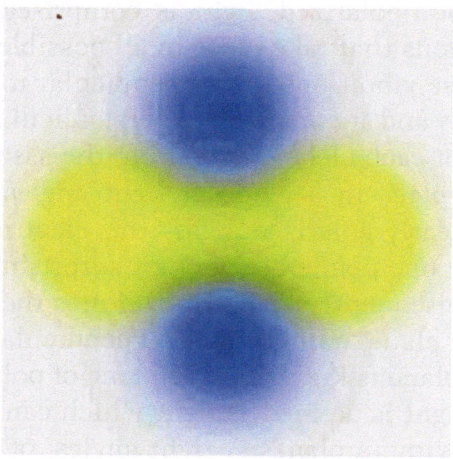

Fig. 6.27. Haidinger's Brush. In reality, the effect is a lot more subtle than depicted here (Credit: Daniel P. B. Smith [original copyright: Daniel P. B. Smith 2006, released under the GFDL])

In 1954, William Shurcliff of the *Polaroid* Corporation found that the brush appears in circularly polarized light as well as in the linearly polarized that Haidinger himself was using. Indeed, it is possible to distinguish between linearly and circularly polarized light by means of Haidinger's brush, and even to tell whether circularly polarized light is left or right handed. For linearly polarized light, the yellow axis of the brush points in a direction perpendicular to the plane of polarization. Horizontal polarization, therefore, results in a vertical brush and this orientation remains unchanged irrespective of how one may incline his or her head. By contrast, circularly polarized light causes the yellow segment to be slanted. The direction of this slant is, once again, independent of how one holds the head, but it differs with respect of the direction of the circular polarization. If we imagine a line bisecting the face, right-handed polarized light results in a yellow brush slanting up to the right of this imaginary line and down to the left of it. For left-handed polarized light, the opposite is true.

It must be appreciated that Haidinger's brush is, literally, in the eye of the beholder (more of this in a little while) and therefore cannot be photographed or instrumentally recorded in any way. Also, because the effect is subtle, the light source requires a good degree of polarized light before it can be perceived. Typically, the degree of polarization needs to be at least 60% for a good chance of seeing the phenomenon. The effect is also only discernible in light toward the blue end of the spectrum (it is missing in polarized red light) and its faintness requires an unpatterned and uniform background.

But it *can* be seen in nature! Sunlight is polarized, with the highest degree at 90° from the Sun. The brush is therefore visible in the daytime sky, and especially near the zenith in very bright twilight. With a background that is blue, uniform and highly polarized, the brush emerges as a faint stain against the sky.

Just as people vary in the sensitivity of their eyesight, so it would seem that there are differences in our ability to detect the polarization of light. Some folk see Haidinger's brush more readily than others and it may be that some of us are not able to detect it at all. Practice, as in most endeavors, undoubtedly helps, and with a combination of innate sensitivity to polarized light and skill acquired over time, some people find it relatively easy to see the brush practically anywhere in the daytime sky. A few have even detected it quite close to the Sun, once the latter is blocked the hand or some other intervening object, even though the degree of polarization is significantly lower there than at large angular distances.

Project 16: Haidinger's Brush

Haidinger's brush is most readily seen in nature near the zenith of a clear sky just at sunset or at sunrise. Make yourself comfortable, and keep your eyes focused on the overhead sky around sunset or (if you are really keen!) sunrise. After 2 or 3 min, the sky will appear to take on a marbled appearance and, shortly thereafter, the brush should emerge as a subtle stain on the sky. Some may find it easily, while for others it might be difficult or the whole exercise may even be a non event. If you are one of those fortunate folk who find this an easy project, you might like to try seeing the brush at smaller angles from the Sun and, should even that be too easy, you may like to try locating Brewster's point. If you succeed at that, congratulations. You have just joined a pretty exclusive club!

These observers can even detect another interesting fact about polarized sunlight at small angular separations. For most of the sky, the brush's yellow branches are radial with respect to the Sun, in other words they point toward it. However, there is a point in the sky known as *Brewster's point* where polarization changes and the yellow branches become tangential to the Sun. Those who are very proficient at observing Haidinger's brush can visually detect this change in orientation and with nothing more than the unaided eye, find Brewster's point in the heavens!

What causes Haidinger's brush? How does the human eye come to perceive this peculiar pattern?

Although the entire story is yet to be told, a partial explanation was given by H. von Helmoltz just 20 years after its discovery. He explained it in terms of *dichroism* (i.e. the exhibiting of different colors in two different directions when viewed in transmitted light) of a pigment within the *retina* of the eye. The region of the retina responsible, however, must include more than just that part – known as the *fovea* – that is responsible for seeing images with higher resolution. The image of the brush is simply too large to be confined to the fovea and must therefore have its origin within the surrounding *macular* area of the retina. Moreover, because the brush is colored, and the perception of color involves the *cones* and not the *rods*, it must be the former that are involved in its perception. As noted earlier, Haidinger's brush is not visible in red light; a region of the spectrum to which the macular pigment is transparent.

The probable dichroic pigment responsible is *Lutein*. This is a long chain molecule that absorbs more polarized light whose electric vector is parallel to the molecule's axis and less light whose electric vector is perpendicular to it. However, the Lutein molecules must also be aligned in such a way that the overall effect is not cancelled out. The alignment required to do the trick takes the form of concentric circles of Lutein molecules in the macular region surrounding the fovea. Imagine a pattern of concentric circles as the face of a clock. Suppose a beam of vertically polarized light strikes the face of this "clock", the sector around 12 o'clock will be the region containing molecules whose axes will, on average, be perpendicular to the electric vector and will therefore absorb little of the light. On the other hand, the sector perpendicular to this – centered around the "quarter hour" on the clock face where the axes of Leutein molecules are mostly parallel to the light's electric vector – will absorb much of it. This would give rise to the characteristic Haidinger's brush pattern, in this example, the "12 o'clock/6 o'clock" axis forming the blue segment and that perpendicular to this displaying yellow.

Because macular pigment is found in a region containing radially arranged nerve fibers – being sandwiched in between the outer and inner limiting membranes of the retina – the abovementioned alignment of a portion of these molecules might be expected. Nevertheless, the percentage of molecules actually aligned in this way will still be quite small, hence the weakness of the Haidinger's brush effect.

Although this arrangement of macular pigment molecules is enough to give rise to an effect roughly similar to Haidinger's brush, it fails to fully account for all the details and present thinking does not believe it to tell the whole story. Other factors (involving certain features of the cones?) are apparently at work as well. After more than 160 years, the mysterious brush has yet to yield all its secrets!

There Is a Large Moon Rising

We have now looked at all manner of clouds, some "regular" and others more or less mysterious, mirages, odd events like dark days and widespread atmospheric opacity and some of the odd and interesting effects that go with this, and last but by no means least, the rarely recognized optical phenomenon of Haidinger's brush. But no excursion into the byways of sky phenomena would be complete without taking a look at an illusion so common as to have been experienced by nearly every soul who has ever lived, yet one which is still often misunderstood and whose "obvious" explanation is, plainly, wrong!

This is the so-called *Moon illusion*; the apparent enlargement of the image of the Moon as seen rising or setting compared with that of the Moon when riding high in the sky. The illusion is not so obvious with the Sun, probably because of that object's great brilliance, but is very striking indeed in the case of the Moon near its full phase.

The "obvious" explanation for this phenomenon is the atmosphere. After all, we have seen in our look at mirages how many tricks our planet's blanket of gas can play, so surely this is just another one. When the Moon lies close to the horizon, it shines through a greater layer of air than when it beams down on us from directly overhead, so it seems natural to suppose that this greater depth of atmosphere acts as a lens and magnifies the image. This looks like a perfectly logical and obvious explanation. The only problem is, it is entirely wrong! For once, Earth's restless envelope is *not* responsible!

The incorrectness of this apparently "common sense" explanation is easily demonstrated by viewing the Moon through a hollow tube just wide enough to accommodate the Moon's image. If the Moon's image truly is magnified when close to the horizon, a tube whose aperture barely fits the image of the Moon at high elevation will not cover it at low or, conversely, if the tube is barely wide enough to fit the rising Moon, the image of the Moon later in the night will be noticeably smaller than the field of view through the tube.

As it happens (see Project 17), this is not so. If the Moon is viewed through the same tube on the same night (the Moon's apparent size varies a little at different parts of its orbit and between lunations thanks to its varying distance from Earth), the size of the image remains unchanged, irrespective of whether the

Moon is near horizon or zenith. That is why we speak of the "Moon *illusion.*" The image only *seems* to be larger near the horizon. In actual fact, it is no different to any other point on the Moon's path across the sky. But if magnification by the atmosphere was the cause of the phenomenon, the image really *would* be larger near the horizon. After all, that is what "magnification" means!

Although it might not be as strikingly obvious, the illusion also affects the constellations. A good example is Orion, the Hunter or as we in the Southern lands less romantically call it "The Saucepan" (its main stars do form a remarkably saucepan like configuration from south of the equator. The appearance is lost somewhat for northern observers, thanks to the different orientation of the pattern). When this spectacular constellation is rising, it looks enormous. When high in the sky, it definitely seems to have shrunk. The Moon illusion becomes the Orion illusion!

So, if atmospheric magnification is not the solution, what is?

We are still not completely sure. The size of an object within our field of view can mean two separate things. It can mean its angular dimension or its actual, physical, dimension. Two objects having the same physical size can look vastly dissimilar if placed at different distances from the eye and, conversely, two objects at different distances can have the same angular size, even though their physical dimensions may be vastly dissimilar. In relation to the Moon illusion, the problem to be solved is whether the Moon at low altitude appears to be further away than the Moon at high elevation (and therefore more distant) or whether it seems to have greater angular dimension.

As long ago as the year 200 AD, Cleomedes proposed that when the Moon is near the horizon, it appears further away and this tricks the mind into imagining it to be physically larger. Experience teaches that objects within the atmosphere that are seen nearer the horizon are generally further away than those seen higher in the sky. This affects the way in which we perceive the sky. We may speak poetically about the "dome of the heavens" but we do not really see it as a true hemispherical "dome." It is more like a dome that has been squashed at the top. We see the sky more like a great tent with steep sides and a rather extensive and more or less flat ceiling. Objects moving toward the horizon (flying birds, for example) move away from us along this gently curving surface. As they recede, their images on the retinas of our eyes grow smaller.

However, the Moon, Sun and constellations, when near the horizon, subtend the same angular image on our retinas as when located higher in the sky, tricking us into thinking that they have somehow grown larger. In effect, the entire region of the sky at lower elevations appears to be stretched out compared to the greater region at higher altitudes. What we call the Moon illusion is really just one instance of a larger illusion which might better be called the "sky illusion."

This hypothesis is now the most popular one, but it does have its critics. The problem is that only very few people (about 5% or thereabouts) claim that the low altitude Moon looks both larger and further away. Most say that it looks larger and *closer*.

An alternative explanation is the so-called relative size hypothesis, or the thesis that the perceived size of an object does not simply depend upon the size of the retinal image, but also upon that of objects in its immediate neighborhood. High in the sky, the Moon (or Sun or constellations) appear in relatively empty regions; empty, at least, of large objects capable of showing definite detail. When near the horizon, the opposite is the case, and it is this fact (or so it is argued) that generates the illusion of a larger Moon.

According to this hypothesis, the Moon illusion becomes a special case of the *Ebbinghaus illusion*. This is best demonstrated by looking at two sets of circles, as in Fig. 6.28. In one set, a central circle is surrounded by a rather spacious ring of large ones. A second set looks more crowded, with the central subject surrounded rather more closely by a ring of smaller circles. The illusion involves the apparent perceived size of the central circle. In the first and more spacious set, the central circle appears smaller than the one in the crowded second set. In fact the diameter of the central circle is the same in each.

A similar suggestion notes that when the Moon is low, common foreground objects (trees and the like) also appear in the same field of view and these tend to give the observer a point of reference. It is suggested that the Moon appears larger if viewed within a field containing some such points of reference. Higher in the sky, similar reference points do not normally exist and the Moon seems to be smaller.

However it may be explained, this illusion again reminds us that, in our strange skies, seeing is *not* always believing!

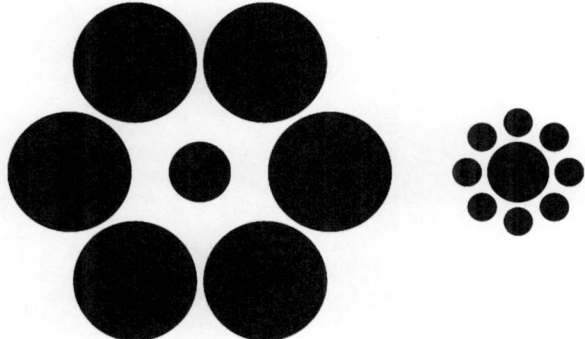

Fig. 6.28. The Ebbinghaus Illusion. Do both solid central circles have the same diameter? Measure them and find out (you may be surprised!)

Project 17: The Moon Illusion

There are several ways of showing that the image of the rising Moon is really no larger than that of the overhead Moon. Try looking at the rising Moon through a thin tube just large enough for its image to fit the field of view. Later that same night, look at it again through the same tube and see whether it continues to fill the field of view.

Another way is to cut a small disk that almost but not quite covers the rising Moon if held at arm's length, (i.e. a thin rim remains visible around the edge of the disk). Does a thin rim remain when the Moon is high in the sky?

Does the Moon seem larger when seen rising (or setting) over a bare horizon (such as the ocean or open prairie) or over a forested or city horizon? Does the nature of the horizon make any difference to the illusion? What is your own experience in this respect?

When the Moon is observed high in the sky but in the same field of view as nearby trees or high buildings, does it still appear smaller than when seen near the horizon? What does this have to say about explanations of the illusion involving the presence of foreground objects?

7. Strange Sights and Sounds of Air and Sky

The sky is, on the whole, a reasonably quiet and serene place. Most days all one hears is the call of an occasional bird in flight, the drone of a distant aircraft or the gentle sighing of the wind.

But there are times when the dome above our heads presents a different face. Crashing thunder, the drumming of rain, the rumbling of hail or the more frightening roar of that monster of the weather systems, the tornado, assaults our ears.

Yet, even these sometimes chilling voices of the sky and air are part of the corpus of accepted experience. There is nothing especially strange about them. There may still be unanswered questions – why can thunder sometimes be heard to unusual distances? Why are tornadoes *so* noisy? – but no deep mystery surrounds these sounds.

But that is not the case with every atmospheric noise.

Things That Go Bump in the Air!

I Hear the Sound of Distant Guns!

From regions as diverse as Ireland, Australia and various parts of North America, people have long reported hearing sounds resembling the discharge of heavy artillery at great distances, variously naming the sounds "Cannons of Barisal", "mistpoeffeurs (fog pistols)," "Marinas," "Moodus Noises," etc. Often, though by no means always, the locations where the phantom guns are heard are near large bodies of water. Lake Seneca, near Syracuse in New York is one such location. Lough Neagh in Ireland is another. Similar sounds have also been heard in the Philippines, where they appear to be in some way associated with the ocean.

But ocean is a long way from Central Australia and the phantom guns have been heard there as well, right from the earliest days of exploration. Early explorers specifically noted that real guns or anything else that might be capable of making such sounds were hundreds – more likely thousands – of miles away!

Writing in his account of the exploration of inland Australia spanning the years 1828–1833, Captain Sturt noted that on 7 February 1829 in a place near the present town of Bourke "about 3 P.M. ... Mr. Hume and I were occupied tracing the chart upon the ground. The day had been remarkably fine, not a cloud was there in the heavens, nor a breath of air to be felt. On a sudden we heard what seemed to be the report of a gun fired at the distance of between 5 and 6 miles. It was not the hollow sound of an earthly explosion, or the sharp cracking noise of falling timber, but in every way resembled a discharge of a heavy piece of ordnance." Sturt continued that "no one was certain whence the sound proceeded" and remarked that nothing unusual could be seen, even by one of the party who climbed a tree for a wider view. The incident, he confessed "is a matter of mystery to me."

Sturt's phantom gun seems to have discharged only once. Frequently however, the Barisal Guns fire rounds of two or three shots, only to fall silent until the next round of phantom ammunition is supplied.

No satisfactory answer has been given for this phenomenon, although there have been any number of suggestions, some clearly more likely than others. The noises are most frequently heard in very calm conditions, but that may only mean that in other circumstances they are drowned out by extraneous noise or stifled by the wind. Calm air may not necessarily be essential for their occurrence.

Certain species of noise-making fish have been put forward as an explanation, but these could (at best) only account for sounds heard at sea or very close to large bodies of water.

Another suggestion that apparently is quite popular as an explanation for the sounds heard in the Philippines identifies the booms as sound made by large waves crashing on the shoreline. Under certain weather conditions, sound waves generated by the surf might be reflected from atmospheric inversion layers or even from clouds and be audible over surprisingly large distances.

Inland regions, where one would not normally expect to hear the surf, might experience the sound of the largest waves as isolated cannon-like booms. And, as surf-riders and those living by the ocean know, large waves frequently come in "sets" of three, just like many reported instances of Barisal Guns.

Explosive sounds can easily be generated when air is compressed by a wave breaking into a cave and blowing back from the mouth of the cavern. If a hole or fissure is present in the wall or roof of the cave, compressed air and water may roar out from this with tremendous and even explosive force in a spectacular fountain of spray. Such blowholes can be exceptionally noisy and the sound not infrequently takes the form of a loud "boom" which fairly shakes the surrounding ground. Under the right conditions, the explosive sounds of a large blowhole might carry as far, or maybe even further, than those of ordinary surf.

Nevertheless, no crashing surf or booming blowholes can explain the sound heard in the middle of the Australian desert by Sturt, Hume and company. A rock cracking in the heat of the summer sun might sound like a sharp pistol shot under the right circumstances, but such could hardly be mistaken for "a heavy piece of ordnance."

Single, isolated, explosive sounds may at times be traced to the sonic booms of large meteors entering Earth's atmosphere. Under overcast daytime skies, even a large meteor may pass unseen though not unheard. Yet, Sturt expressly remarks on the meteorological conditions at the time of his experience and notes that the day was clear and cloudless. It is difficult to imagine that a meteor large enough to generate the sound he reported would have passed unnoticed by at least one member of the party, even in broad daylight. Moreover, even had the meteor itself been somehow missed, there would very probably have been a persistent smoke trail in the sky which is not likely to have remained unseen by a party scrutinizing the surroundings for the source of a mystery sound.

Earthquake sounds have also been suggested as the firing mechanism of the phantom guns, however no convincing correlation with seismic disturbances has been forthcoming.

Another suggestion is bubbles of gas escaping from rotting vegetable matter on the floors of lakes and other bodies of water. As we will see in a little while, this mechanism can indeed be implicated in some anomalous sound cases, but it is stretching credibility to extend this explanation to all (or possibly even most) reports of phantom guns.

Bizarre "explanations" such as flying saucers breaking the sound barrier have also been mooted, but for obvious reasons will not be dignified with further discussion!

A point which is not clear is the distance of the sound's source. The traditional Barisal Guns always *seem* distant. But is this apparent remoteness genuine? Or are they a little like the rainbow that always remains out of reach, appearing to recede faster than an observer can approach it? In other words, is the apparent remoteness of the Cannons of Barisal illusory?

Occasionally however, a very loud explosive sound for which there is no obvious explanation *is* heard. Is this a Barisal Cannon firing near at hand, a related phenomenon or something altogether unrelated to the apparently distant booming? I don't know the answer, but I do know from personal experience that these loud reports can be very startling indeed!

What Was That?!

The year was either 1991 or 1992 and the writer was standing in a store in his home town deep in conversation with a friend when all of a sudden there occurred the most horrendous "bang!" Everybody in shops and in the street was startled. "What was that?" became the question of the moment, though not always expressed quite so blandly!

As far as I am aware, that question remains unanswered to this very day. Suggestions were legion, but each failed at some point to do justice to what happened.

An isolated bolt of lightning may superficially have appeared the most plausible explanation. The skies, as I recall, were partially cloudy although the clouds were not of a type that would normally be expected to discharge lightning bolts. Moreover, nobody actually saw a flash and it was later learned that a lightning detector, owned by a resident of the area, failed to register any discharge at that time. As more information became available, it was clear that the sound was audible over a very wide area; far larger than that normally expected for a thunderclap.

Other possibilities such as sonic booms from aircraft, a meteorite or re-entering space debris were no more successful. No aircraft was heard or seen and no meteor or burning space debris reported.

As the region has a number of coal mines servicing several nearby large electricity generators, long-wall underground blasting in one of the mines was also suggested. Once again, the extended area over which the report was heard would seem to rule this out.

In the days following the sound, a couple of odd accounts reached the writer's ears. A young lady told me that her dog became strangely agitated immediately before the sound. The young lady concerned had been a casual acquaintance of mine for several years and I have no doubt as to her honesty.

The second odd account is one that cannot be personally verified by the author as it was passed down by word of mouth (at least two mouths actually). This is not intended as a criticism, as there is no reason to doubt the story's veracity. Apparently, at the time of the sound, a "micro-tsunami" – a small isolated wave – raced across the otherwise glassy surface of a nearby estuarine bay. The exact details are not known to me, but the witness was apparently left in no doubt that the two events were intimately associated.

It is possible that the cause of the dog's agitation was something unrelated to the noise, although I find that unlikely. The wave's being a mere coincidence seems even less likely. Both events might appear to hint at some seismic disturbance, although nothing was registered on seismographs at the Riverview Observatory in the northern suburbs of Sydney.

Actually, more than one mystery bang was heard from Sydney and surrounding areas in the early 1990s, none of which has been explained.

Curiously, many years earlier – sometime in the late 1950s – another very loud report rocked the region. This happened on a fine and clear day free from any chance of lightning. It occurred at a time when sonic booms from aircraft at a nearby Air Force base were not uncommon (fortunately, they are no longer permitted to exceed Mach 1 over populated areas) but that does not appear to have been the cause of this event.

Is She About to Blow?

The region of south-western Victoria and south-eastern South Australia was the last area of Australia to experience active volcanism. Stories handed down by generations of Koori people tell of mountains spitting fire and the "legends" even trace the progression of active volcanic vents as continental drift caused the land to slowly slide over hot spots creating new volcanoes even as old ones became extinct.

One of the last to close down Vulcan's forge was Tower Hill in Victoria. Some estimates have this still spitting and belching early in the Common Era, although others think it closed down activity somewhat earlier than this. In any case, the low flat-top hill continues to be graced by a large caldera, one corner of which is occupied by a crater lake. The floor of the caldera is now a reserve for wildlife and the settlement of Tower Hill overlooks this from the top of the caldera cliffs. Several years ago my wife, Meg, and I enjoyed the unusual experience of being at eye level with a majestic eagle cruising back and forth along the edge of the cliffs.

Although Tower Hill has not been extinct for very long by volcanic standards, geologists assure us that these Australian volcanoes are drifting further and further away from the "hot spots" and are very unlikely to rejuvenate, Vesuvius like, in any foreseeable future. But the residents of Tower Hill may have had their doubts about this one early morning several years ago when loud explosive sounds were heard coming from the lake at the bottom of the Tower Hill caldera. Was Tower Hill township about to become Australia's Pompey!?

Fortunately, no. Investigations found that the explosions were not evidence that the volcano was coming back to life again. The real culprit was decaying vegetable matter on the lake's floor. Apparently, gas generated from this material rose to the surface in large bubbles which burst with loud reports, echoing through the caldera in the early morning quiet.

The source of these mystery explosive sounds was not hard to track down. Similar events happening in remote ponds or swamps may not be as easily found and may well be the cause of *some* reports of phantom gunfire.

The Skies Quake

On the morning of April 4, 2006, residents of San Diego experienced what some thought was an earth tremor. Yet, no earthquake was recorded at that time.

Not everyone agreed that it felt *exactly* like an earthquake however. Writing on his weblog *brianstorms*, Brian Dear commented that "the windows of the building shook, the door shook, the walls shook, but not the floor, not my chair, not the table. Right at the moment it was happening, I thought sonic boom – but it went on too long."

Explanation?

No real explanation, just a statement that it must have been a "skyquake", whatever that is supposed to be.

Now, if you would like a really weird tale, try this next one.

Back in 1977, a self-styled psychic made some "predictions" for the years ahead. Like most statements of this type, they were a mixed bag. Rock' n' roll king Elvis Presley would die that year (he did!), the (Australian) Labor Party would win the forthcoming election (it didn't) and, after coming to power, would remain for several terms (it did that eventually, but only following a later election). The *piece de resistance* of the predictions was to be a landing of alien spacecraft around the year 2000 which, needless to say, did not happen, but in the midst of this mixed bag of "predictions," came the announcement that the Australian capital city of Canberra would very soon experience an earthquake. Remarkably, just days after this was published in a New Age type magazine (though I'm not sure that this term was even in use then!), residents of Canberra were surprised by a gentle but definite shaking. News media immediately reported it as a mild earthquake and the psychic's supporters were elated. But their confidence was premature. When seismographs were examined, no evidence of an earthquake was recorded. Whatever it was that shook Canberra, it was evidently *not* an earthquake.

There are several suggestions as to what these so-called skyquakes might be. Sonic booms have been proposed, but appear unlikely as a general explanation, although they might explain a few instances. One variation of this suggestion has the phenomenon resulting from the sonic booms of top secret hypersonic aircraft being developed by the U.S. Air Force. As reports have been found dating back to at least the 1700s, no more need be said about this one.

The shifting of the continental shelf has also been suggested, but (if it is capable of explaining any instances) this can hardly account for something that is reported as far from the coastline as Canberra. Another suggestion involves a volume of air suddenly becoming hotter than the surrounding atmosphere and exploding like a balloon (I'm not sure what to make of that one!).

Large exploding meteors once again provide a tempting suggestion and it is true that artificial Earth satellites report explosions in the kiloton range happening several times each year on the edge of outer space as large and fragile bodies explosively disrupt upon entering the atmosphere. The difficulty with this explanation lies however, with the fact that for a meteoroid to penetrate far enough into the air to produce effects at ground level, it surely would be widely observed as a bright fireball. Those obvious meteoric events that have been noted as producing loud sounds and vibrations at ground level have also been very bright and conspicuous, leaving no doubt as to their nature.

Things That Go Bright in the Night

Strange sounds of uncertain origin are not the only mysteries of Earth and air. Every bit as puzzling are the unidentified lights which are reported with surprising frequency, even by astronomers, pilots and others who are familiar with the recognized denizens of the nocturnal landscape.

These mystery lights have a long history and have served as the source of much superstitious lore and many a folk tale over the years. From the corpse candle to the flying saucer, human imagination more than human reason has often come to the fore in the quest for understanding of these strange glowing denizens of the dark.

Although the division is not entirely clear cut, nocturnal lights can be roughly separated into two broad classifications; those that occur close to or even on the ground and those that are observed in the sky itself. This division, we repeat, is far from absolute, but it is the one that we shall follow – for convenience sake if for no other reason – in the following pages as we go hunting these elusive apparitions.

Low-Level Nocturnal Lights; Spooky Lights Near the Ground

Around the world, stories abound of small spooky lights appearing and vanishing at, or close to, ground level. Called by various names – *jack-o'-lanterns*, ignis fatuus (foolish fire), *corpse candles*, *fairy lights*, *wills-o'-the wisp*, *ghost lights*, *spook lights*, and so forth, it is sometimes difficult to see where folk lore ends and reality begins. It is also not always apparent that these terms relate to the same phenomenon (or even to related phenomena, for that matter), a point that will become clearer later.

Certain places appear to be unusually productive of these weird luminosities. Where this happens, the "local" variety of light is frequently given a special name. Thus we have the *Waimea Light* of Hawaii, the *Marfa Ghost Light* of Texas, *Brown Mountain Light* of North Carolina, *Naga Ghost Fireballs* of Thailand and so forth.

Reading reports of these events as folklore, one can easily become dismissive of their reality. "Maybe," we are tempted to think, "The people who see these lights were unusually prone to letting their imaginations run wild and merely overreacted to a perfectly common source of light. After all, it is known that people have even reported the rising Moon as a UFO!" This line of reasoning sounds very plausible, and is almost certainly accurate in certain instances. But when we, or close friends or relatives whose testimony we have always relied upon, experience something truly bizarre, the whole thing becomes a different ball game.

Allow me, at this point, to relate an experience which happened to two such reliable relatives of mine back in 1951. The witnesses were father and daughter and they were traveling by road northward through the western country of England to meet a friend at Carlisle in the wee small hours of the morning. The friend was arriving by rail sometime before dawn and they were planning to drive all night to be in Carlisle to meet him.

At some point late in their journey, the father and daughter noticed a single orange-red light far off to one side of the road. At first they thought it may have been the fire under the boiler of a steam locomotive and wondered if, perhaps, this was the one carrying their friend. They quickly realized, however, that the light was not in the right direction for the railway line.

Driving a little farther, they became aware that the light was moving closer to the road in the direction toward which they were headed. As the path of their vehicle and that of the mystery light began to converge, the shape and form of the "light" became clearer. It assumed the form of a single tongue of flame – at least 10 ft in length – with its long axis parallel to the ground. It appeared to be floating roughly 10 ft above the surface and looked rather like the exhaust flame of a rocket; but without the rocket!

The "flame" (as we may now call it) eventually passed over the road ahead – stretching from one side to the other – and (to the daughter's horror) stopped! She pleaded with her father not to drive underneath it but he, despite obvious fear, opted instead for pushing down the accelerator and speeding beneath the strange apparition. His daughter crouched to the floor until they were safe on the other side. Emerging, she glanced back through the rear vision window to see the "flame" apparently reversing away from the road; back in the direction from whence it came!

Despite enquiries by the father from both townsfolk and academics, no explanation was found. It had scant resemblance to the flickering evanescence of the will-o'-the-wisp. Astronomical explanations are also ruled out for obvious reasons. If it was a mirage, it was unlike any other!

The only explanation that I can think of is a dense swarm of some type of luminous insect, but even that seems pretty tenuous. It seems that this one will remain a mystery unless some reader has had a similar experience and managed to find a satisfactory answer.

The Min Min Light of Central Australia

Needless to say, Australia has its fair share of ghost lights and, simply because this is the writer's home country, we will exemplify the phenomenon through looking at some of these.

Australian ghost lights have, apparently, long been known to the indigenous Koori people of this land who, in common with native inhabitants the world over, have their own histories and folk tales about the phenomenon. The Koori believe these lights hale from the spirit world. They are "dead men's camp fires" according to some; "debil-debils" (ghosts) according to others. The picture of the spirits of past generations of Kooris gathered around spectral campfires in the desert, relating stories of how the land was long, long ago has a certain romantic appeal and is one which many a modern-day Koori might wish to believe.

The most famous – or maybe we should say "infamous" – of Australian spook lights is the *Min Min Light* which allegedly haunts the Boulia district of western Queensland. The light is so named because it traditionally appears around the ruins of an old hotel and mail exchange once known as the Min Min pub, located on the boundary of the two large Warenda and Lucknow stations. By the way, in Australia a "station" does not necessarily mean a railway station. When used of properties (as in the present instance) it is similar to what North Americans would call a "ranch." Warenda and Lucknow are "ranches"; ranches on a grand scale. Some of the property holdings in central Australia are truly enormous. Before subdivision into smaller stations, one exceeded the area of Great Britain. Warenda and Lucknow were not as large as that, but the location of the old Min Min pub at the boundary of such extensive properties meant that it was extremely isolated. The region where the Light was most frequently reported – a stretch of about 40 miles between Min Min and Hamilton River – is an expansive tract of almost treeless country once described as "inhospitable to the extent that only [someone who knows the area] may imagine."

Nevertheless, in a country where the largest private properties exceed the smallest European countries and where "down the road a bit," when given as a direction by a local, can mean a distance of 300 miles or more (that is not an exaggeration – the writer has a specific instance in mind!), the "region around Min Min" can swell to include an area larger than Texas. It should not be surprising therefore that "Min Min Light" has over time become essentially synonymous with any spook light of central Australia. We will return to this later when attempting to evaluate the status of this strange phenomenon.

Fig. 7.1. The Min Min Light is a tourist attraction around Boulia in "outback" Australia (Credit: GondwanaGirl, 6 January 2009)

A classic sighting of the light, insofar as any sighting can be deemed "classic" or "typical", is provided by Henry Lamond in a 1937 issue of *Walkabout* magazine.

The incident happened many years previously, during the winter of 1912. At the time, Mr. Lamond was managing the Warenda station (then the largest in Queensland) which he described as "being a few miles short of the round 5,000 [miles]" though he failed to mention whether this was in area or perimeter. Either way, it covered a lot of country!

On the night in question, he had to leave the main house at Warenda and travel by horseback to a sort of out-station known as Slasher's Creek to start lamb marking. Slasher's Creek was about "12 miles below Min Min."

Because of office work at the station headquarters, Lamond did not begin his journey until 2 A.M., anticipating arrival at the out-station just before daylight.

After riding between 6 and 10 miles out onto the "high downs" (essentially extensive grasslands used for grazing) he noticed what he initially thought was a headlight of a motor vehicle. Motor vehicles, he noted, were not common in those days and, although it was not *very* rare to encounter one, it was something that he did not expect in that location at such an hour of the morning.

It was soon apparent, however, that this was no motor vehicle headlight. He writes that as the light drew closer;

> It remained in one bulbous ball instead of dividing into the two head-lights ... it was too greeny-glary for an acetylene light; it floated too high for any car; there was something eerie about it.

The horse became agitated, snorting as the light approached. Lamond continues;

> The light came on, floating as airily as a bubble, moving with comparative slowness ... I should estimate now that it was moving at about 10 mph and anything from five to ten feet above the ground ... Its size, I should say, ... would have been about that of a new-risen moon.
>
> That light and I passed each other, going in opposite direction. I kept an eye on it while it was passing, and I'd say it was about two hundred yards off when it suddenly just faded and died away.

Lamond remarked that he put forward no theories as to what the light may have been, disagreeing with an unnamed friend who suggested that night birds with luminous fungus on their feathers offered the best explanation (a popular, though unsupported, theory of spook lights). Notwithstanding his professed lack of commitment to an explanation, Lamond nevertheless remarked that "the area, for miles up and down the [Hamilton] river, is a region of mud springs – things from which strange scents and musty gases flow as well as water."

On the face of it, this is a striking sighting. Yet, there is one niggling problem. The *Walkabout* account was published on that day when many folk find it strangely acceptable to tell lies, tall stories and generally engage in other forms of behavior ruled inappropriate through the rest of the year. I refer, of course, to April 1, April Fool's Day!

Of course, odd things can truly happen, or be reported, on this day as on any other. I recall the time in 1986 when the announcement of observations at Siding Spring Observatory confirming complex and biologically interesting organic material on Halley's Comet, was posted on the Observatory's notice board on April 1. Some of the staff needed a deal of convincing that this was not a hoax!

The niggling doubt that Lamond's story might have been a hoax is, however, largely cleared away by a second account of the incident which he gave many years later to the *North Australian Monthly* and published in January 1961. Nevertheless, this clarification came at the cost of another problem. The second account is almost identical with the first, except for a small inconsistency in his description of the behavior of the horse he was riding. In the earlier account, as noted above, Lamond clearly said that the horse snorted and became agitated. By contrast, according to his later retelling, the horse showed no reaction to the light. This was noted by Lamond as an indication that the horse did not mistake the light for a headlamp, as "had it been a car coming toward us, that mare would have been afraid." One version must be wrong, at least with respect to the detail of the horse's reaction. What this says about the accuracy of the rest of the account is anybody's guess.

Fortunately, the next report that we shall consider has greater consistency. It is also far more recent, having happened on May 2 in the year 1981. The witness this time was Detective Sergeant Lyall Booth of the Police Stock Investigation Squad at Cloncurry in Queensland and is quoted in the *Royal Geographical Society of Australia Bulletin* by no less than N. W. Bauer, then the Queensland Commissioner for Police.

According to the report, Booth was camped on the bank of what is known as the Bulla Bulla waterhole, about 60 km (around 37 miles) east of Boulia. The only other person present at the time was the camp cook who resided a short distance north of where Booth spent the night.

The night was cool and clear, and Booth dropped off to sleep around 9 P.M. but awoke some 2 h later for no apparent reason. At that time he, "saw a light which at first I took to be a car headlight ... I thought it was a vehicle on the main road, but after a short time I realized that the main road was further to the north and that vehicle lights could not be seen plainly from my location. The light appeared to be just to the west of the Hamilton River channels and appeared to be moving but it did not seem to get any closer."

According to Booth, the light was lower than tree-top level and appeared to fluctuate (which, he opined, may have given it the illusion of movement). It remained a single light source, white in color "similar to the light thrown by a quartz iodine headlight."

Booth kept the light under view for about an hour, before drifting off to sleep. Awaking again about 1 A.M., he noticed the light (or, *a* light) just north of where the cook was camped. If this was the same one that he spied earlier, it would have moved about a kilometer to the south-west of its initial location. This light may, however, have been a different one, as it appeared fainter and shone with more of a yellow color. Nevertheless, it was still bright enough to illuminate the surrounding ground and was judged by Booth to be about 3–6 ft above the surface, moving a short distance from west to east and then remaining stationary. After about 5 or 6 min, the light appeared to dive toward the ground and disappear.

Booth was convinced that this light did not have a human source, mainly because of the remote and lonely location in which it was seen. He mentioned that the night was clear, but an astronomical explanation is improbable. At that time of year, the bright star Arcturus rises in the north-east, but by the time of night that Booth's sightings were made, it would have climbed far too high in the northern skies to be confused with anything near ground level. The relatively stationary appearance and long duration of the light likewise rules out other astronomical explanations, including bright artificial satellites or head on meteors appearing at very small elevation.

Lack of electrical activity plus long duration also excludes ball lightning, unless some of its forms are even stranger than even the most bizarre reports imply! On the face of it, the light(s) observed by Sergeant Booth seem to have been genuinely mysterious.

But all may not be as it seems! The Booth sighting has been more recently re-examined by scientist Bill Chalker, a long time investigator of anomalous phenomena. Chalker notes that, although the main road to which Booth referred certainly headed in a direction that made the sighting of headlights difficult, detailed maps of the area show a secondary road running approximately along the line of sight connecting Booth's camp with the apparent position of the first mystery light. Not only that, but the same line of sight also passes through the Hamilton River Hotel, albeit some 20 km (just over 12 miles) away.

It is possible that the first light at least may have been headlights (although admittedly, the duration was rather long) or even a bright lamp being used by someone camping close to a parked vehicle somewhere along the secondary road. It is unlikely that lights from the hotel would in normal circumstances be so brightly visible over these distances, but Booth's remark that the night was "cool" may give a clue to another mechanism at work here. Cool air close to the ground can refract light rays from a distant source to create a nocturnal mirage, as discussed below. This may have been what Booth saw between 11 P.M. and midnight.

The second light is more difficult to explain in this way. Indeed, if this light was not really a "second" one at all, but a continuation of the "first", the explanations suggested here probably fall down completely.

A Ball of Worms?

Chalker suggests that the second light may have been a swarm of luminous insects. You may recall that this possible explanation was offered earlier (albeit as an act of desperation!) for the strange "drifting flame" seen near the Scottish/English border in 1951.

Actually, there is some evidence that swarms of luminous insects do occur in the Min Min region. A journalist named James Oram describes an interesting conversation with a Mr. Allan Camm who claimed to have experienced a very odd encounter with a Min Min light whilst driving a road grader at night along an almost straight stretch of road between Boulia and Dajarra. Somewhere along this road, Mr. Camm became aware that a luminous object was following him. At this, he turned his vehicle around and drove straight at it! In his own words "It was flat ground and I got the grader up to its top speed of about 25 miles/h. I drove straight into it. It exploded all around me and then I found its secret. The light was made up of millions of glow-worms. In fact, when I got to the pub at Dajarra, I still had the worms in my hair and clothes."

This is a most interesting account, but it is unfortunate that Allan Camm has eluded Chalker's attempts to track him down. There is no implication here that the story is not true, just that it remains second hand and, as recounted, may omit some potentially important details.

Even if the original witness could be located and this second-hand account verified beyond all shadow of doubt, the mystery of the Min Min light may still not have been completely solved. Chalker points out that, in certain respects, the Camm incident is atypical of most accounts of the Min Min light. Most notably, the glowing ball did not switch off as Camm's grader approached. In most accounts, the light goes out if a person (on foot, let alone driving a piece of heavy road working equipment!) comes too close. Other accounts suggest that the light is a little like a rainbow in that it always remains out of reach. Walking – or driving – up to and actually impacting it is clearly not typical.

On the other hand, the Camm account compares nicely with a minority of spook-light stories from other parts of the world.

For example, one Professor J. Castell-Evans wrote in the Welsh *Daily News* newspaper on February 15, 1905 about an incident that happened to him as a youth. Apparently, there was a certain location alleged to have been haunted by a ghost that took the form of a globe of light and, young men being what they are, a number of them (as a dare no doubt) tried on occasions to catch it. Most of the time, the light kept out of reach in true spook-light fashion. Eventually, Castell-Evans succeeded in persuading his braver friends to join him in forming a circle around the light and trapping it. He writes that "When we got close enough I put out my hand. The thing was cold. It gave me a nasty shudder. It broke into little bits as I touched it. Then we all bolted. But the ghost was nothing but phosphorescent insects."

Remarkably, this Welsh chemistry professor encountered another spook-light in his youth, near Bela Lake in the valley of Dee. This time, he pelted it with a stone, which passed right through it and disrupted it into "a thousand little pieces" which after a few moments "joined together again; and once more the ball of light went dancing down over the course of the stream. It was nothing but a cluster of luminescent insects."

An odd event that occurred in the writer's own region in the mid-1970s may also have a similar explanation. A group of young people camped overnight on a nearby beach were startled when they witnessed a glowing orange ball about the size of a football emerge from some low bushes just beyond the edge of the sand. As the ball rose into the air, numerous glowing worm-like objects poured out of it and swarmed all around the (now terrified) witnesses.

It may or may not be a coincidence that, on another occasion, a motorist driving alone the coastal road quite close to this location reported seeing a luminous orange ball float out of the vegetation and drift across the road in front of him, rising higher as it moved. This object (which, by the witness' description, was quite a deal larger than a football) was also said to have shed smaller luminous objects along the way.

Swarms of luminous insects would appear to fit the description of both objects, although I know of nothing indigenous to the region that glows with an orange light or forms such dense swarms. Certain types of luminous fungus have been reported in the wider area, but these can hardly account for moving lights. Similarly, beetles of the *Lampyridae* family are quite common in nearby regions, although I do not personally know of any having been seen close to the site of the abovementioned incidents. In any case, the light these insects emit is not of an orange hue. Known as *fireflies* in Australia, they give bright bursts of bluish-white light but, unlike the tropical species, do not gather in swarms giving the impressive synchronized light displays reported from these regions.

On several occasions, the writer has noted small and very bright blue-green "sparks" in the household lawn. Upon investigation, the culprit turned out to be a small species of earthworm. One time, whilst engaged in comet sweeping with a pair of tripod mounted binoculars, I reached for my star chart only to find the hard cover of the book of charts crossed by a brilliant luminous stripe! Once again, a small earthworm was responsible. It had slithered across the cover apparently leaving a trail of luminous slime that kept lighting up brilliantly when disturbed. It would fade, then light up again when rubbed, repeating this several times before gradually fading away.

The true source of this illumination is not the earthworm itself, but certain luminous bacteria inside its gut. Presumably, they are quite common in the area, but it is difficult to understand how they could account for floating globes of orange light. As I am not an entomologist however, I am not in a position to exclude the possibility of other local forms of bioluminescence that may be responsible.

Whatever the status of the bioluminescence explanation for these local sightings, there seems little doubt that at least a subclass of spook-lights are balls of luminous insects. Camm's light appears to have been one of these. Maybe Booth's light (at least, his "second" light) might have been another. But are all Min Min lights – and spook lights in general – open to this explanation?

Let's look at some more reports and further assess this possibility.

More Lights: More Explanations?

During his summer vacation in 1972, a university student found work plowing at night on a remote property close to the western New South Wales and Queensland border.

About 1:30 in the morning of December 16, he decided to stop the tractor for a short break and soon became aware of a noise like that of an approaching high powered engine. Despite a clear view of surrounding open country, nothing could be seen (no headlamps or the like) even though the apparent source of the sound seemed to pass by quite close at hand. Gradually, the sound died away and the student paid it no further thought as he resumed his plowing.

Then, several minutes later, a light appeared. It entered the paddock being ploughed and moved toward the tractor, floating above the ground at a height of about 10 ft. As it approached, the student could see it more clearly, noting that it appeared as a small spherical body with a brighter concentration of light toward its centre. Nevertheless, the greater share of the light seemed to radiate from the surface of the sphere, not from the concentrated central region. This illumination was bright enough to light up a wide area of ground.

Some 50 yards from the tractor, the object made a smooth turn and then stopped, hovering for about 5 s before vanishing ... *only to reappear outside the paddock fence in a direction behind the tractor!*

It would be possible to give an account of this light in terms of luminous insects, were it not for this final "trick." Unless there were two swarms – one lighting up as the other switched off – some alternative explanation appears called for here. The same must also be said for the following incidents as well.

Professor Jack Pettigrew of the University of Queensland studied reports of the Min Min light for several years and has seen it himself on more than one occasion. The first strange light he saw looked like the planet Venus setting. The only problem; it simply refused to set. It just sat on the western horizon for an unusually long duration!

On a later occasion, while driving with two colleagues through western Queensland, he saw what at first appeared to be the eye shine of a cat in front of the vehicle. Apparently suspicious that this was more than a pair of cat's eyes, the three stopped the car and turned off its headlights. But the "cat" remained visible, bobbing around as if it had a will of its own.

The light, whatever it was, *appeared* to be just a short distance down the road. But was this apparent proximity real? The three companions had sharply different opinions about that.

To settle the argument, Pettigrew drove the automobile off road and across the surrounding plains, using its compass to work out the light's parallax. Remarkably, he had to drive about 5 km before any discernible change in the light's position became noticeable. From this, he calculated its distance at some 300 km (about 188 miles); well beyond the horizon!

A Min Min Mirage?

This great distance gave Pettigrew an important clue as to the light's true nature. He concluded that it was probably a nocturnal version of the type of mirage known as Fata Morgana about which we spoke in the previous chapter. We might recall that, where a temperature inversion exists close to the ground (that is to say, where the air immediately above the ground is colder and therefore denser than that at slightly greater altitude), light from a source close to the ground can be refracted in such a way as to travel in a path following the curvature of the Earth's surface. This enables the light source to be visible far beyond the normal range. If the light rays simply traveled in straight lines, with no significant refraction, their source's area of visibility would be greatly reduced. In effect, the dense and cold layer of air acts a little like an optic fiber, allowing the light to be transmitted for hundreds of miles without being dispersed.

As it happened, Pettigrew subsequently learned that a distant motor vehicle had indeed been heading in his direction at the time the light was seen. Readers in more populated areas than the "outback" of western Queensland may have difficulty in getting their minds around a situation where the only two vehicles on a stretch of road were separated by several hundreds of miles, but let it be said that some of the roads in the Australian desert are literally the ones less traveled! On a more somber note, a mechanical breakdown in such regions can be a death sentence unless one is well prepared for unexpected contingencies. Min Min light hunting is not for the faint hearted!

A Home Grown Min Min Light!

In order to test his hypothesis, Professor Pettigrew decided to create his very own Min Min light!

Choosing a night when the weather conditions appeared just right for a low level temperature inversion (namely, a cool night following a hot day with little breeze) he drove his vehicle about 6 miles over a slight rise and into a dry watercourse. Back at the starting point of his journey, six observers watched the progress of his vehicle's headlights. From their position, the headlights should have passed out of the line of sight, yet that did not happen. Instead, the observers saw the light from Pettigrew's vehicle float above the horizon, just like the phantom "cat's eyes" of his earlier sighting.

As an added proof that the professor was on the right track, dawn the following morning brought forth a typical diurnal Fata Morgana of a distant range of hills. The range would normally not have been visible, being over the horizon, yet "floated up off the horizon and gradually got dissected by fingers of blue sky" before eventually sinking out of sight as the Sun warmed the air and caused the temperature inversion to disperse.

This phenomenon may also explain Pettigrew's earlier observation of a Venus-like luminary that refused to set. The setting object might well have been this planet, but the light that lingered on the horizon was probably a mirage of Venus caused by a low level temperature inversion. We have already encountered the odd phenomenon that causes astronomical objects below the horizon to be visible as mirages floating just above it. This is the Novaya Zemlya effect, also discussed in the previous chapter. Some researchers of the Min Min light, and spook lights in general, see in the Novaya Zemlya effect a possible explanation for many of these odd reports.

Is the Mystery Solved?

It is indeed tempting to think that the mystery of the Min Min light, and by extension, all similar spook lights in different parts of the world, has largely been solved. Luminous insects account for a few, but the more important contribution is made by nocturnal mirages. Throw in the occasional case of genuine swamp gas and ... mystery solved!

Well, maybe.

As already mentioned, Bill Chalker suggested that the first light sighted by Booth may be explicable in terms of a mirage of the nocturnal Fata Morgana variety. In support of this suggestion, he draws attention to astronomer Eric Silverberg's hypothesis that the Marfa light of Texas – a spook light phenomenon similar in many ways to the Min Min – results from mirages of motor vehicle headlights carried over great distances "along writhing paths by atmospheric tunneling" due to temperature inversions "bending light in funny ways." In effect, Silverberg's explanation for the Marfa light corresponds to Pettigrew's of the Min Min.

Of course, vehicular headlights cannot account for very old Min Min and other spook light sightings, but the same mirage effect also works for bright stars and planets near the horizon, as suggested by Pettigrew's non-setting Venus incident. Also, mirages of the distant camp fires of Kooris may have given rise to the phenomena in earlier days. The Koori explanation for these lights may be partially correct; except that the camp fires were not the ectoplasmic flames kindled by ancestral spirits, but physical fires warming living contemporaries.

But what are we to make of stories like the one involving the university student plowing after midnight?

Had his location not been so remote and the hour so late, one might suspect a prank was being played by his campus friends, but a lonely paddock at 1 o'clock in the morning is a little extreme even for university students. In any case, half the fun of such a practical joke is revealing it to the "victim" at a later time and watching his reaction; something which apparently did not happen.

Part of the problem is trying to determine the accuracy of the student's description of the "light." Remember that this happened late at night. Was his description influenced by tiredness, general fatigue, fear or the tendency toward disorientation that can so easily happen in the darkness? His description appeared quite detailed and non-sensational, but was his estimate of the light's distance and movement accurate? We simply cannot know.

The light's behavior of switching off in front of the tractor and immediately appearing behind it possibly hints at a mirage interpretation, but then the observer's estimate of the light's proximity would be wrong and it seems difficult to account for his statement that it illuminated the ground.

And then there is the noise! Although there is no strict reason for connecting the noise and the light, it does seem strange that two seemingly anomalous events (one auditory and the other visual) should have occurred at the same place within minutes of one another if they were completely unassociated. But the association need not have been direct. It might simply have been that the meteorological conditions suitable for the occurrence of one phenomenon also favored the occurrence of the other.

Sound waves, like light rays, can also be refracted by temperature inversions. In the oceans, whales communicate with one another over amazing distances by exploiting the way sound travels through water layers of differing density. Sounds can travel long distances in the air by, first of all, radiating skyward and subsequently being refracted downward after bouncing off inversion layers. Some areas – distant from its source – will hear the sound while others (possibly even closer to the source) will not. As conditions change, so the areas capable of hearing the sound will also change.

Just for fun, let's try the following as a possible explanation for this "noisy" Min Min light sighting. On the night in question, conditions were ideal both for long-range sound refraction and nocturnal Fata Morgana type mirages. Somewhere quite distant from the paddock being ploughed, a large overnight transport lorry rumbled along the road. Maybe its headlights were not facing in the right direction to trigger a Fata Morgana mirage observable by the student or, maybe, something else prevented him from observing this as he ploughed the paddock. Thanks to an appropriately placed temperature inversion, the noise of the truck's engines bounced off the layer and returned to ground level far beyond the distance over which such a vehicle could normally be heard. The sound took a relatively long time to reach the student (the exact length, of course, depending upon how distant the truck really was) but fortuitously timed its arrival for when the student was taking a break from his task. The difference in the velocities of light and sound are so great that, even had a mirage of the truck's lights been formed, it would already have vanished by the time the sound arrived. Hence the "invisibility" of the sound's source.

The observed light may then have been a mirage of another vehicle (not necessarily one as large and noisy as a truck) coincidentally traveling along the same distant road, possibly in the opposite direction from that of the first and therefore in a better position to form a luminous mirage at the relevant location. The sound may have arrived some time later, but was not heard as the tractor motor had by then been restarted.

As we said, this little exercise is not to be taken too seriously. There is no evidence known to this author as to the presence of a road in the requisite direction, still less of a large truck was traveling it at the time. Moreover, it ignores the difficulties that we have already seen attending the "mirage" hypothesis as an explanation for this case. If the above little frolic is indeed a credible explanation for this Min Min sighting, certain details in the report would need to be less than accurate. This is not said in criticism of the witness. If I encountered something that strange, alone in a remote location at the dead of night, I'm sure my description of what was seen would have plenty of inaccuracies! On the other hand, the above may prove less of an explanation for this sighting than an example of just how superficially plausible any "explanation" of anything can be made to sound, especially if certain aspects of the original report are conveniently *assumed* to be inaccurate!

In any final analysis of the Min Min Light – in so far as there can be such a thing as a "final" analysis of something of this nature – we would probably be wise to agree with Bill Chalker in seeing the "Light" as much as a social phenomenon as anything else. The Min Min is a legend of country Australia. The region beginning near the old Min Min ruins and spreading outward to cover much of the country has, over the years, acquired the reputation of being "haunted" by the Light and this reputation has itself been strengthened by reports of strange luminous phenomena. Any light that appears to be even slightly anomalous tends to be credited as another sighting of the Min Min Light, thereby gaining sustenance from the legend while at the same time adding its own contribution to the legend's continuation. Stories told in far western hotel bars – told in ways that best entertain weary travelers, thirsty cattle drovers and hard-working shearers not necessarily according to the rules of strict scientific accuracy! – flow into the veins of the legend and are preserved somewhere in that limbo between pure folk lore and established fact.

Swarms of luminous insects, nocturnal mirages and probably a mix of quite "ordinary" light sources such as distant headlights (without benefit of Fata Morgana), distant aircraft and who-knows-what-else all contribute to the legend of the Min Min light.

Enough has been discovered to confirm some interesting phenomena (the nocturnal Fata Morgana for instance) and it is not impossible that some even stranger phenomenon lurks amongst the years of reports, but essentially the Min Min is a legend and it is this legend that binds together the various diverse luminous phenomena which, in their turn, keep it alive. I suspect that this is the same for every spook light "haunting" where ever in the world these may be located.

Flickering Lights on Marshes and Rivers

Spook lights such as Min Min and their counterparts in other lands are not the only low-level luminous phenomena. Even more famous are the transient flickering lights sometimes seen over marshes and swamps. Known, as we remarked earlier, by a variety of names, of which *will-o'-the-wisp* is probably the best known, these lights have long been the stuff of folklore. Although there is no hard borderline between this phenomenon and spook lights of the Min Min variety, it is probably best to treat the flame like will-o'-the-wisp type sightings as distinct from the glowing balls typical of spook lights. If we are right about the diverse nature of the latter, we should certainly treat the former as a separate category with a very different cause.

The flickering lights have long been explained as swamp gas. Although the exact mechanisms are still far from completely understood, it seems that gases rising from decaying organic matter in marshes or stagnant bodies of water can under certain conditions spontaneously combust and produce brief flickering flames. In many instances, the "flames" are apparently of unusually low temperature and do not set fire to surrounding dry grass. On other occasions however, fires have been reported as having been started by spontaneously igniting swamp gas.

A variety of gases can be released from rotting organic material. Methane may well have acquired the title of "marsh gas", but hydrogen sulfide can also be an unwelcome product of marshes. The rather playful name of "rotten egg gas" (beloved of schoolboys' making stink bombs) somewhat belies the dangerous nature of this substance. Not only is it inflammable, but it has the curious property of being odorless in the high concentrations in which it is also

Fig. 7.2. Faux-o-the-wisp. A piece of burning birch bark floating on a length of wood was used to make this false will-o'-the-wisp. Color of the flame has been altered (Credit: Tuohirulla)

dangerously toxic. The rotten egg smell only betrays its presence at small and harmless levels. It is when you can't smell it, that it is time to be worried!

That hydrogen sulfide is sometimes responsible for spontaneous flames is implied by an amazing display witnessed in Blundellsands, England, on the evening of June 5, 1902. In the words of one witness,

> In the distance we first saw smoke with frequent jets of fire bursting forth from the mud of a shallow canal. Drawing near, we perceived a strong sulfurous odor, and saw little flames of fire and heard a hissing sound as though a large quantity of phosphorus was being ignited. … The phenomenon lasted some time, until the tide covered the part and quenched the fire … As we returned from our walk the atmosphere was impregnated with a strong odor of sulfur.

The mention here of a strong sulfurous odor probably hints at the presence of hydrogen sulfide, though well diluted as it escaped into the surrounding air.

Mekong Fireballs … and a Perfect Murder?

This unpleasant gas, or one of similar nature, may also be responsible for one of the more spectacular and recurring displays of spontaneous flames – the so-called *Naga Fireballs* of the Mekong River in Thailand and Laos.

This phenomenon is most active during October each year. At that time, just after the end of the rains, balls of light rise from the Mekong River and drift high into the air. Measuring roughly the size of hens' eggs, they are said to be of a reddish color and rise hundreds of feet into the atmosphere before eventually fading out. On some nights, several tens of lights are seen. On other nights, they can number in the thousands!

Villagers along the river believe that the glowing balls are the work of the Naga or Phaya Nage, a snake said to inhabit the river, but scientists have proposed a somewhat more prosaic, if no less interesting, mechanism. Fermentation of organic sediment on the bed of the river, not a mythological serpent, is believed to be the real snake in the grass. Gases rise in bubbles to the surface and spontaneously combust upon reaching the air.

The phenomenon of gases rising from fermenting organic matter has also, we may recall, been implicated in the occurrence of some mystery explosive sounds, as in the Tower Hill incident in western Victoria. But in 2006, a documentary film maker uncovered evidence that this process might provide the answer to yet another mystery. A suspected double homicide that has baffled law enforcement and the curious public alike since New Year's Day of 1963!

We refer to the famous Bogle-Chandler case; one of Australia's most intriguing and best known murder mysteries.

On the eve of the New Year, physicist Dr. Gilbert Bogle, and colleague Mrs. Margaret Chandler were enjoying a party at Chatswood in the northern suburbs of Sydney. Mrs. Chandler's husband Geoffrey, also a scientist employed, like Dr. Bogle, at the *Commonwealth Scientific and Industrial Research Organisation* (CSIRO), briefly called in at the party early in the night, before going off to another with friends belonging to a movement known as the "Push." This could best be described as a counter-cultural and more or less libertine group of artists and intellectuals that grew up in the 1950s. We can be sure that the party Geoffrey Chandler attended later in the night was not involved in drinking lemonade and talking football!

Neither was the Chatswood party it would seem. All the folk involved had what might euphemistically be called "modern" views about relationships; a point of considerable importance to those investigating the incident.

In essence, what transpired was that soon after 4 A.M., Dr. Bogle left the party to drive Mrs. Chandler to her home. That was the last time either was seen alive. Come morning, the partially clad corpses of both were found by the car in a grove of trees on the banks of the Lane Cove River.

No toxins were found in the bodies, neither was there any sign of injury nor anything that could have resulted in death.

Of course, speculation ran riot. Geoffrey Chandler was suspected. Bogle had a well deserved reputation for liking the ladies (to put it mildly) and therefore jealousy became the suspected motive. Yet, Chandler had a sound alibi and, in any case, would a follower of the Push and an apparent believer in open marriage be upset over such a bourgeois matter as a fling between his wife and one of his best friends; even had such a liaison taken place? (For which, it must be said, there was absolutely no evidence).

Other theories fared no better. Nobody other than Geoffrey Chandler was known to have had the slightest motive to murder the pair, and if he didn't do it all avenues simply came to dead ends. There was a suspicion of experimentation with drugs such as LSD (not then widely known ... 1962–3 was still the pre-hippy era!) and an accidental overdose of this substance was suspected. This avenue of inquiry also ended at the proverbial brick wall.

Then, in 2006, the most plausible theory to date was put forward. The murderer was none other than the Lane Cove River itself and the murder weapon, hydrogen sulfide. This theory was first made public that year in a television documentary *Who Killed Dr. Bogle and Mrs. Chandler* by filmmaker Peter Butt.

Back in 1963, the Lane Cove River was very polluted (I, for one, would not like to swim in it even today!) and not far from where the bodies of Bogle and Chandler were found, a factory had once stood. In an era less aware of the damage done by dumping organic wastes into rivers and streams, the effluent from this factory went straight into the water and settled to the bottom. Over the years, it decayed and (so the theory goes) gave off bubbles of hydrogen sulfide. On certain nights – and there is reason to believe that the night of December 31, 1962–January 1, 1963 was one – meteorological conditions were such that a relatively concentrated layer of this gas could accumulate over the river surface and spread out onto the surrounding banks. Concentrations could become high enough for the "rotten egg" smell to disappear and any unsuspecting person deciding to stretch out on the banks of the river to (let us say) "watch the stars" might find themselves overcome before they realized what was happening.

Come the morning breeze and the gas would be gone, leaving only its victims of the night before. The river had committed the perfect murder!

High-Level Nocturnal Lights: Spooky Lights in the Sky

Not everything that goes bright in the night does so close to ground level. Over the years, an impressive array of reports concerning luminous phenomena at high altitude has also accumulated and some of this continues to be quite baffling. Gathered around this subject, perhaps more than any other, is a wealth of speculation, supposition and straight-out dogmatism that has both clouded the issue and given the topic such a less-than-respectable aura as to easily scare away the serious investigator. It is truly amazing how readily some people accept that a simple point of light in the sky is an extraterrestrial spaceship. Take it from one who has investigated a number of alleged sightings!

Very often, the reported light has a well-known explanation. I recall one occasion where the light turned out to be our old friend Venus seen at low altitude and probably interrupted by some small clouds. It was identified with 100% certainty, but even if it had not been, there were still plenty of things that could have given a similar appearance. Aircraft lights, other bright stars or planets, balloons or kites with lights attached, to name a few. Yet, when asked what she thought the light was, the witness immediately replied "a spaceship." From the context, it was obvious that she did not mean one launched in USA or Russia.

In addition to these types of preconceived beliefs, it is also extremely difficult to get a fair estimate of distance at night, and if distance is grossly over – or under – estimated, the size and true brightness of the light source suffers accordingly. Is that light drifting overhead a firefly leisurely coasting along at tree-top level or a large body traveling at supersonic speed on the edge of outer space?

Sometimes even a very familiar object can look strange at night. On a couple of occasions, I have seen faint lights passing high overhead at speeds which were too slow for meteors, yet the objects themselves looked too "soft" (almost, though not quite, describable as "diffuse") to be artificial satellites. Only after catching one of these objects in a pair of binoculars did its true nature become obvious. They were simply high flying night birds catching the glow of urban lights. The oiliness of birds' wings makes them quite good reflectors.

Mad Meteors

Meteors not infrequently give rise to high-level nocturnal light reports. The typical meteor appears as a star-like object rapidly moving for a short distance across the sky before either fading away or terminating in a bright flash of light. Some sport transitory tails, others leave behind them persistent trains of slowly fading luminescence. A minority takes the form of very bright balls of light and a rare few even appear nebulous or cloud-like.

Some, however, defy expectations. Wavering and even curved trajectories have been reported from time to time. There have even been reports of a meteor apparently falling toward the horizon executing an about-face and "bouncing" back upward again!

Because odd and erratic meteors have been discussed at some length in my book *Weird Astronomy* little need be said here, although any talk of high level nocturnal lights should at least give them a mention.

The curved trajectories of certain meteors are sometimes explained as optical illusions. Noting that many curved meteors have been reported as initially visible on the periphery of vision, the suggestion has been made that a quick turning of the head (once the meteor has been detected) for a better view gives the illusion of a curved trajectory for the object. This may well be so in some instances, but the explanation wears a little thin for those reports where the meteor was always seen more or less at the centre of the field of vision.

Sharp changes in direction, even including apparent reversals, are very difficult to explain if taken at face value. The usual suggestion is that the original body fragments; the parent meteor quickly fading out while the "child" continues on, albeit along a somewhat different path. To the observer, it looks as though a single meteor has abruptly changed direction. This explanation might have a somewhat contrived appearance, but what are the alternatives?

Of course, if an apparent meteor appears too mad and erratic, perhaps it is time to suggest that it is not really a meteor at all.

Other Strange Lights in the Night Sky

It is indeed apparent that not all high level nocturnal lights can be explained away as meteors, however odd. Something that stays in one place for a significant length of time is not easily explained in that way. True, a head-on or "point" meteor having a slow velocity may remain as a single stationary light for a few moments, but even the duration of the slowest of meteors is still brief. And what can be said about the curious property of some reported lights flashing out at one point only to instantaneously reappear at another? This suggests a mirage-like phenomenon although the meteorological conditions at the time of each sighting would need to be assessed for such a conclusion to be convincing.

But is this explanation capable of accounting for the likes of the following?

On February 10, 1951, Mr. C. Rhodes was camped in outback Queensland, not far from the scant remains of the Min Min Hotel. Consequently, the light that he reported seeing around 8.30 that night is sometimes listed as a Min Min light, although it bears little resemblance to the low-level spook lights discussed earlier and is not easily accommodated to the explanations suggested for these.

The light first appeared in the northern sky and initially remained stationary. After a little while however, it began to move and streaked swiftly westward for about 40°, then paused, "jumped up and down" before once more coming to rest. It then disappeared behind (or maybe into) a small cloud, remaining briefly hidden from sight before reappearing again at the cloud's bottom. Simultaneously, a second light appeared at the top of the cloud. Both then started moving in a westerly direction, but fused once more into a single light before clearing the cloud. The duration of the entire event was estimated as about 10 s.

This account bears some resemblance to a curious sighting made by the writer and two other people from Lane Cove in Sydney's northern suburbs back in 1972. A ridge of hills (more correctly, the escarpment of a low plateau) runs along the northern parts of Sydney suburbia and from there one can enjoy an expansive vista over the Sydney basin. At night, an ocean of lights spreads out to the far horizon. We were looking across this expanse when attention was drawn to two very bright lights a little above the horizon. They were white and brilliant, far exceeding any star or planet. Actually, if they were as distant as they appeared, their luminosity must have rivaled that of incoming international aircraft with all lights blazing as they made their decent into Mascot Airport some distance to the left. However, unlike the aircraft, these lights were stationary. They were at an oblique angle to one another, such that an imaginary line drawn through them would have subtended an angle of about 45° with the horizon.

As we watched (for several seconds), the lower of the two lights began to flicker and, at exactly the same time, a third flickered into existence above what had formerly been the uppermost of the original two. The oblique alignment remained the same. In other words, the same imaginary line drawn obliquely through the first two passed through the third as well. Moreover, the distance between the first two lights equaled (as far as we could determine) that between the remaining member of the first pair and the newcomer.

As if that were not strange enough, what happened next left us open mouthed. The newest light appeared to "jump" from its original position to the right of the lower light (i.e., what had been the uppermost of the original pair) to a position 45° to the left of it. It didn't move across the sky to its new position. Really, it just blinked out and (another?) instantaneously blinked on at the new position. This new orientation remained constant for several seconds before the new light started flickering and faded out. The remaining one may have sunk a little toward the horizon and appeared to pulsate a little, though the former may have been simply an optical illusion and the pulsating was no more than that expected for a light source shining through a dense layer of atmosphere over a large city. Our attention was momentarily distracted and when we looked back, this remaining light had also disappeared.

What were they?

Frankly, I don't have a clue. Perhaps it was a mirage, although I would need a lot of convincing of that. If they were significantly closer than they looked, they may have been artificial lights strung on kites or toy balloons, but once again I find that explanation unconvincing. Let me be quite clear however that I do not believe that anything "paranormal" was causing the lights. They were not spaceships from another world, or anything of that nature. On the contrary, I am quite sure that their explanation lies within the field of the commonplace. I just don't know what that explanation might be.

That is probably the best conclusion to the issue of high level nocturnal lights in general. After all, it is very difficult to discern the nature of something seen against a night-time sky. Increasingly, we must contend not only with natural phenomena but with artificial ones as well. Air craft (including secret military ones) and artificial satellites are just two classes of things that we are putting into the air and into space. Seen under unusual circumstances, even something as common as an aircraft can look pretty weird.

Yet, as remains true of low-level nocturnal lights, so here too, the possibility of rare and as-yet unknown phenomena may truly account for a minority of reports. This is why we should retain a degree of skepticism about any conviction that everything can be explained "rationally" (which, in the context, actually means "in terms of what is already known"). If it turns out that nothing appears adequate to explain an event, we should not force an explanation in terms of something known. That is not "rational." In fact, admitting that some unknown phenomenon might be responsible is the more "rational" approach in that situation. An "unknown" phenomenon certainly does not imply anything non-rational. It was not long ago that ball lighting was not an officially acknowledged phenomenon and sightings of this were squeezed with increasing difficulty into "acceptable" explanations. Even more recently the electrophonic sounds of meteors became respectable, thanks largely to the work of Professor Colin Keay, and we no longer need go to the sometimes absurd ends of trying to explain observations of this effect in terms of psychology. Maybe the residue of high level nocturnal lights that we continue to find difficult to accommodate to existing explanations is really telling us that something still needs to be discovered. But when and if it is, I'll wager that it has no connection with little green men from outer space!

Ironically however, just because the folk belief associating high-level nocturnal lights with extraterrestrial visitation has grown up over the years, the mere mention of this class of phenomenon has, in a sense, brought us full circle. We began this journey through our strange skies with thoughts about our place in the wider universe and here we are at the end of this book once again looking skyward and wondering. Although I disagree that extraterrestrial visitors play any role in the sightings of high-level nocturnal lights, the contrary belief has at least had the effect of turning the thoughts of many to the wider universe and our place therein. I guess it is fitting therefore, that having delved right down to Earth with low-level nocturnal lights, we now cast our glance skyward again to embrace the same view with which we began; the broad and fascinating vista of our strange skies.

Index

D.A.J. Seargent, *Weird Weather: Tales of Astronomical and Atmospheric
Anomalies*, Astronomers' Universe, DOI 10.1007/978-1-4614-3070-4,
© Springer Science+Business Media New York 2012